BIRDS, BATS, AND BLOOMS

THEODORE H. FLEMING

BIRDS, BATS, AND BLOOMS

The Coevolution of Vertebrate Pollinators and Their Plants

THE UNIVERSITY OF ARIZONA PRESS
TUCSON

The University of Arizona Press
www.uapress.arizona.edu

We respectfully acknowledge the University of Arizona is on the land and territories of Indigenous peoples. Today, Arizona is home to twenty-two federally recognized tribes, with Tucson being home to the O'odham and the Yaqui. Committed to diversity and inclusion, the University strives to build sustainable relationships with sovereign Native Nations and Indigenous communities through education offerings, partnerships, and community service.

ISBN-13: 978-0-8165-5372-3 (paperback)
ISBN-13: 978-0-8165-5373-0 (ebook)

Cover design by Leigh McDonald
Cover photos of Dark long-tongued bat and Emerald hummingbird by Theodore H. Fleming, photo of *Hamelia patens* from AdobeStock.
Designed and typeset by Leigh McDonald in Arno Pro 11/14 and Bulmer MT (display)

Library of Congress Cataloging-in-Publication Data
Names: Fleming, Theodore H., author.
Title: Birds, bats, and blooms : the coevolution of vertebrate pollinators and their plants / Theodore H. Fleming.
Description: Tucson : University of Arizona Press, 2024. | Includes bibliographical references and index.
Identifiers: LCCN 2024000589 (print) | LCCN 2024000590 (ebook) | ISBN 9780816553723 (paperback) | ISBN 9780816553730 (ebook)
Subjects: LCSH: Plant-pollinator relationships. | Birds—Evolution. | Bats—Evolution. | Plants—Evolution. | Coevolution.
Classification: LCC QH549.5 .F53 2024 (print) | LCC QH549.5 (ebook) | DDC 577.8—dc23/eng/20240618
LC record available at https://lccn.loc.gov/2024000589
LC ebook record available at https://lccn.loc.gov/2024000590

Printed in the United States of America
♾ This paper meets the requirements of ANSI/NISO Z39.42-1992 (Permanence of Paper).

CONTENTS

PREFACE

In her song "Woodstock," Joni Mitchell sings "We are stardust . . ."—a common theme that appears in many songs. And it's literally true. If you travel back far enough in time, everything in the universe started out as stardust. We're all a form of cosmic dust that was ejected from precursors of stars and that eventually condensed to form our solar system, atoms, and us.

In this book, I will concentrate on what I think are some exceptionally interesting forms of stardust: nectar-feeding birds and bats and their food plants. In doing so, I will address two big questions dealing with the evolution of life on Earth: How did these animals evolve, and how did they coevolve with their food plants? I've long had an interest in these questions because of my research experiences as a field biologist working in the new-world tropics and the Sonoran Desert and, just as importantly, where I currently live.

As an obvious example of the importance of location, three species of hummingbirds reside year-round in my yard in northwest Tucson, Arizona. This allows me to watch their behavior every day. These birds include Anna's, broadbill, and Costa's hummingbirds (please see the glossary for scientific names of all birds, bats, and plants mentioned in this book). Each of their adult males are easily distinguished. In full sunlight Anna's males have a brilliant red crown and a short red throat gorget (elongated throat feathers); broadbill males are dark green with a bright-orange bill and lack a gorget; and Costa's males have a brilliant purple crown and a long purple throat gorget. Females and juveniles of these species all lack gorgets and hence are far less distinctive and require close observation to identify them.

In addition to visiting our three hummingbird feeders year-round, in the spring these birds feed at a variety of tubular-shaped native flowers in our backyard: orange *Justicias*, red or purple *Salvias*, and red or pink *Penstemons*, and red nontubular fairy dusters. They also visit non-native tubular flowers, including several species of red or yellow aloes and red cape honeysuckles. They do not visit the nontubular yellow or orange flowers of desert marigolds, brittlebush, paloverdes, mesquites, and Mexican poppies; they also ignore the white flowers of daisy fleabanes and deep blue flowers of desert bluebells. Flower shape and color are obviously very important for food choice in these fascinating birds.

Lucky for me because I'm a specialist in bat biology, in the fall our hummingbird feeders are also visited at night for about two months by a nectar-feeding bat, the lesser long-nosed bat, prior to its migration back to Mexico. Its usual food at this time of the year comes from the night-blooming pink or yellow tubular flowers of native paniculate agaves (century plants), but many of them now visit hummingbird feeders throughout southern Arizona. In our neighborhood, they also visit the large bell-shaped white flowers of the non-native Peruvian apple columnar cactus. Like most nocturnal mammals (and birds), these bats are far less gaudy than diurnal hummingbirds, but nonetheless, they are quite handsome. They have relatively large eyes, elongated snouts, fawn-colored fur, and gentle dispositions (in contrast to the very feisty hummers).

Nectar-feeding hummingbirds and bats are common throughout the new-world tropics (or neotropics). Seasonally, several species of migrant hummingbirds, but not nectar bats, are also common throughout western North America as far north as southern Alaska. These beautiful tiny birds are beloved by humans everywhere and are common visitors to hummingbird feeders throughout their geographic ranges. The same cannot be said about nectar-feeding bats, whose geographic ranges barely reach the southwestern United States. There are at least two reasons for this. First, bats are mysterious because of their nocturnal, cryptic lifestyles. And second, at least in the New World, is their long association with their distant relatives, vampire bats. Whereas most people have a strongly positive feeling toward hummingbirds, they tend to have a negative attitude toward bats of all kinds, especially in Latin America. This is a pity because, as we will see, both hummingbirds and nectar bats have played important roles in the reproductive success and evolution of hundreds of species of their food plants in the neotropics. And as a result, the interaction between these nectar-seeking animals and their food plants has resulted in the evolution of a significant fraction of Earth's biological diversity (or biodiversity).

But nectar-feeding birds and bats are not solely restricted in their distributions to the New World. In the old-world tropics and subtropics (or paleotropics) another set of nectar-feeding birds and bats, very distantly related to their neotropical counterparts, visit a somewhat different array of flowers and provide evidence for the widespread phenomenon of evolutionary convergence. Throughout the world, this phenomenon attests to the aphorism "Life imitates life." Despite different geological and biological histories, life in far-flung regions of Earth has come up with amazingly similar responses to many kinds of ecological opportunities. Is this to be expected or is it surprising? Perhaps a detailed examination of the processes behind these evolutionary convergences can help us to begin to answer this question.

Accordingly, in this book I will take a detailed look at the lives and evolutionary histories of these two groups of nectar-feeding vertebrates and their food plants (table 1). In doing so I intend to use these species and their interactions as a model system for understanding the evolutionary processes involved in creating a conspicuous and colorful portion of the diversity of life that we see on Earth today. Because of my research experiences, I will focus mostly on neotropical nectar-feeding birds and bats and their food plants. But a comparison of these neotropical and paleotropical plant-animal interactions is warranted to complete our understanding of certain aspects of the evolution and coevolution of life on Earth.

TABLE 1 Overview of the major families of nectar-feeding birds and bats discussed in this book

ORDER	FAMILY	NO. GENERA	NO. SPECIES	MASS (IN G)	DISTRIBUTION AND APPROXIMATE AGE IN MILLIONS OF YEARS (MA)
Apodiformes (hummingbirds)	Trochilidae	112	363	2–20	New World (22 Ma)
Psittaciformes (parrots and lorikeets)	Psittaculidae, Loriinae (old-world lorikeets and lories)	19	61	20–240	SE Asia, Australasia (10 Ma)

TABLE 1 *continued*

ORDER	FAMILY	NO. GENERA	NO. SPECIES	MASS (IN G)	DISTRIBUTION AND APPROXIMATE AGE IN MILLIONS OF YEARS (MA)
Passeriformes (songbirds)	Meliphagidae (honeyeaters)	53	189	7–244	Africa, Australasia (50 Ma)
	Nectariniidae (sunbirds)	16	145	4–38	Africa, SE Asia, New Guinea (35 Ma)
	Dicaeidae (flowerpeckers)	2	50	5–13	Palearctic, SE Asia, Australasia (35 Ma)
Chiroptera (bats)	Phyllostomidae, part (nectar bats)	19	56	6–30	New-world tropics (20 Ma)
	Pteropodidae, part (flower bats)	6	14	8–82	Old-world tropics (25 Ma)

Note: Minor nectar-eating families not treated here include Drepanididae (Hawaiian honeycreepers), Promeropidae (South African sugarbirds), Thraupidae (new-world tanagers and honeycreepers), Zosteropidae (old-world white-eyes), and Tarsipedidae (Australian honey possums).

BIRDS, BATS, AND BLOOMS

1

BEGINNINGS

ATOMS, EARTH HISTORY, AND THE EVOLUTION OF LIFE

AS FAR AS WE currently know, living organisms of a vast array of different kinds—from tiny bacteria to huge blue whales—occur in only one place in the universe—on planet Earth, whose age has been estimated using radiometric dating to be about 4.54 billion years (Ga) old. Multicellular organisms—the typical plants and animals we see around us today—began to evolve here quite recently, geologically speaking—only about 600 million years ago. And Earth's current assemblage of nectar-feeding birds and bats and their food plants are much younger than this. They only date from the last 50 million years or so—a little over three hundred thousand human generations ago. So the beautiful coevolutionary interactions that I'm about to describe began just a few minutes ago in the history of life on Earth.

Before beginning this story, however, let's take a brief look at what life on Earth is like in the wider context of the universe. Based on our current understanding, stars, galaxies, and planets began to evolve some 13.5 billion years ago as a result of the Big Bang. Shortly after that enormous cosmic event (i.e., within microseconds) the four forces of the apocalypse—the strong and weak subatomic forces, the electromagnetic force, and gravity—began to set in motion the evolution of subatomic particles and eventually atoms and, via the process of nucleosynthesis, elements of increasing atomic size. The two lightest elements—hydrogen (H) and helium (He)—were the first to form as stable atoms and are still the most common elements in the cosmos today. Together they represent 98 percent of our universe's chemical composition. Other familiar elements include oxygen (O, 0.80 percent), carbon (C, 0.36 percent), and nitrogen (N, 0.09 percent)—all major chemical components of life. After sufficient cooling had occurred as the universe expanded, its so-called ordinary matter, which contains much less mass and energy than its so-called dark matter, began to condense to form stars and galaxies that eventually produced elements heavier than He via stellar nucleosynthesis, a process that involves nuclear fusion.

The basic building blocks of all matter are atoms, whose existence and structure were actually unknown in detail until very recently. For example, electrons were discovered in 1897 by Joseph John Thomson; the existence of an atom's nucleus containing positively charged protons in addition to negatively charged electrons was deduced between 1908 and 1913 by Ernest Rutherford and his colleagues; and neutrons with a mass similar to protons but with no electrical charge were discovered in 1932 by James Chadwick. In 1913 the Danish physicist Niels Bohr postulated that atoms consisted of a nucleus surrounded by electrons circling it in discrete orbits, but this model, so familiar to most of us from our school days, was discarded in favor of another one published by Erwin Schrödinger, an Austrian physicist, in 1926. Schrödinger proposed that electrons occur in 3D waveforms or atomic orbital zones separated by quanta of energy rather than in discrete orbits. This model suggests that we should view an atom as containing a nucleus made up of protons and neutrons surrounded by a cloud of tiny electrons.

But as Erwin Schrödinger (1867–1961) wrote in 1944 in his influential book *What Is Life,* detailed physicochemical knowledge of the nature of atoms does not necessarily help us to understand how life evolved here on Earth. This is because a living, respiring, and reproducing organism—even tiny bacteria—is obviously much more than simply a collection of randomly interacting atoms. Whereas individual atoms are in constant motion relative to each other and are always subject to entropy (i.e., the movement toward greater states of disorder through the loss of energy), living organisms exhibit tremendous stability because their vast array of atoms (some seven octillion or 7×10^{27} in humans!) are not free floating. Instead they reside in a vast array of molecules whose atoms are held together by different kinds of chemical bonds. As a result, writes Schrödinger (1944, p. 68), "life seems to be orderly and lawful behavior of matter, not based exclusively on its tendency to go over from order to disorder, but based partly on order that is kept up." In plants, this order comes from the "consumption" of sunlight (electromagnetic energy); in animals, from the consumption of organic matter (chemical energy).

In his 1944 book, Schrödinger was especially interested in the nature of the gene, which he felt had to be a large molecule of tremendous stability because it is usually passed intact from generation to generation with only an occasional mutation (in his words, a "quantum jump" to a new allele). In 1944, the nucleic acid DNA (deoxyribonucleic acid), a member of one of the major classes of chemical compounds found in living organisms, had just been identified as the bearer of genetic information. Schrödinger's speculations about the nature of the gene (he thought it was a kind of protein) were to inspire James Watson and Francis Crick (among others), the codiscoverers of the helical structure of DNA in 1953, to begin to work on this problem. His book is also credited with starting the field of molecular biology, perhaps the predominant domain of much of biology today. As we will see, since the late twentieth century molecular biology has led evolutionary biologists to gain an ever-increasing understanding of the history of life on Earth.

As I've mentioned, planet Earth began to form some 4.5 billion years ago (i.e., after our universe was about two-thirds of its current age) via accretion of particles attracted together via the force of gravity from solar nebulae (i.e., collections of gas and dust along with hydrogen and helium). The solar system to which we belong, with the sun at its center and its planets, moons, and asteroids circling around it in a flat plane, began to form via accretion about 4.6 billion years ago. As we all know, Earth is the third of eight planets from our sun, and its size and distance from the sun have provided especially favorable physical conditions for the evolution of living organisms. The "goldilocks" conditions that support life here include a salubrious temperature in which liquid water can exist year-round and an oxygen-rich atmosphere dense enough in which animals (and airplanes) can fly. Before the evolution of life began at about 3.8 Ga, however, Earth went through a period of Sturm und Drang (i.e., turbulent stress) conditions involving frequent bombardments by asteroids and comets and violent volcanic outgassing from a very thin, hot crust. These machinations created an atmosphere rich in water vapor, carbon dioxide, carbon monoxide, and nitrogen but containing very little oxygen. As our Earth cooled, water vapor in its atmosphere condensed to form bodies of water in which life began to evolve.

How did life, with its current rich mixture of membrane-bound cells containing complex carbon-based (i.e., organic) compounds, such as nucleic acids, proteins, lipids, and carbohydrates, evolve? It must have done so in water through the evolution of simpler compounds such as the precursors of these compounds: nucleotides, amino acids, fatty acids, and simple sugars. The early evolution of life, a totally fascinating and important field of scientific study today, is not really germane to the story I'm about to tell. But some of its major features are worth noting.

Based on chemical evidence, we believe that single-celled organisms had evolved on Earth by 3.8 Ga, only about 0.7 Ga after Earth's formation. Deep-sea hydrothermal vents may have been the sites of their early evolution. The chemical evolution that resulted in the production of the complex compounds found in living organisms today was probably initially catalyzed by early forms of RNA (ribonucleic acid), a more reactive kind of molecule than its sister molecule DNA and one that is capable of catalyzing the formation of amino acids, the basic ingredients of proteins. And for the next 3.2 billion years, life remained single-celled. During this period, two basic forms of single-celled organisms known as archaea and bacteria evolved. Until 1977, both of these kinds of organisms were classified as "bacteria." But detailed study of their genomes, structure, and metabolism supports the idea that they had separate evolutionary origins and represent truly different domains of life. Though they probably had separate evolutionary origins, early on archaea and bacteria likely exchanged genetic information with each other via horizontal (or lateral, within-generation) gene transfer (HGT), a process that continues in single-celled life today. The early lateral (rather than vertical from generation to generation) gene transfers between bacteria and archaea may explain the universality of life's biochemistry and its genetic code. HGT between viruses and bacteria or from one kind of bacteria to another today, for example, is responsible for the rapid development of bacterial resistance to antibiotics. Thus, it is likely, as David Quammen describes in his fascinating book *The Tangled Tree*, that both archaea and bacteria, rather than just one or the other, are ancestral to all other forms of life on Earth. This suggests that the base of the so-called tree of life consists of a web of interconnections between these two kinds of microorganisms rather than one discrete kind of ancestor.

All of life's species can be classified into two major groups based on the condition of their chromosomes: prokaryotes and eukaryotes. The earliest forms of life, including bacteria and archaea, are prokaryotes whose cells lack a membrane-bound nucleus. Their single circular chromosome consisting of DNA and a few proteins is free floating in the cell's cytoplasm. In contrast, the multiple chromosomes and their DNA found in cells of eukaryotes, which include algae, fungi, animals, and plants, reside in a membrane-bound nucleus.

But nuclear DNA is not the only kind of DNA in eukaryotic cells. Also present in their cells are two important kinds of DNA-containing organelles: mitochondria (in both animals and plants) and chloroplasts (only in plants), which represent the presence of bacterial endosymbionts and their DNA that were "captured" long ago by and incorporated into these cells. Mitochondria are important for energy production in eukaryotic cells whereas chloroplasts are responsible for photosynthesis in plants. Genetic analysis has revealed that some of the genetic material found in the nuclei of eukaryotes originated via HGT from the genomes of their endosymbionts. In addition, these analyses have indicated that about 8 percent of our human genome has been acquired via HGT from retroviruses that have become endogenous in some of our cells and tissues. Retroviruses owe their name to the fact that they use their RNA genome to make DNA (the opposite of what normally occurs) that can then be incorporated into an organism's nuclear DNA. The most infamous retrovirus is HIV-1, which causes AIDS in humans and other primates. Because of the pervasiveness of HGT, many molecular biologists have suggested that we should now view life in its many forms, including us, as mosaics of the gene products of many very distantly related kinds of organisms. This reminds me of the saying "It takes a village" but in regard to forming a eukaryotic organism.

Finally, to complete the major events that have led up to the evolution of multicellular life, beginning about 0.6 billion years ago, we need to consider the evolution of Earth's atmosphere from an oxygen-deficient state at its formation to its current oxygen-rich state of about 21 percent of its chemical composition. The source of this oxygen, of course, comes from photosynthesis in which chloroplast-bearing organisms (certain bacteria and green plants) produce it as a byproduct during the conversion of sunlight into chemical energy (e.g., carbohydrates). So when did photosynthesis first evolve? It had to first occur in microorganisms, and of these, cyanobacteria (formerly known as blue-green algae) were the most likely "inventors" of oxygen-producing photosynthesis. They are known to have been common in oceans by about 2.7–2.5 Ga. Shortly thereafter (geologically speaking), by about 2.3–2.1 Ga, oxygen concentrations in the ocean and atmosphere began to increase slowly. By about 0.5 Ga it had reached over 50 percent of its current atmospheric value. After a peak of about 35 percent atmospheric oxygen at about 0.25 Ga in the Triassic period, it decreased to its current value. I'll be referring to the geological timescale frequently throughout this book, and much of it is summarized in table 2.

TABLE 2 The geologic timescale based on the Geological Society of America, version 5.0

ERA	PERIOD	EPOCH	AGE AT BEGINNING OF EACH EPOCH (MA)
Cenozoic	Quaternary	Pleistocene and Holocene	1.8
	Tertiary-Neogene	Pliocene	5.3
		Miocene	23.0
	Tertiary-Paleogene	Oligocene	33.9
		Eocene	56.0
		Paleocene	66.0
Mesozoic	Cretaceous	Late	100.5
		Early	145.0
	Jurassic	Late	163.5
		Middle	174.1
		Early	201.3
	Triassic	Late	237.0
		Middle	247.2
		Early	251.9
Paleozoic	Permian	3 named epochs	298.9
	Carboniferous	7 named epochs	358.9
	Devonian	Late	382.7
		Middle	393.3
		Early	419.2
	Silurian	Several named epochs	443.8
	Ordovician	Late	458.4
		Middle	470.0
		Early	485.4
	Cambrian	10 named epochs	541.0

Note: Ages are given in millions of years (Ma).

High levels of oceanic and atmospheric oxygen were likely one of the most important conditions necessary for the evolution of multicellular life on Earth. Among other things, relatively high oxygen levels set the stage for aerobic, as opposed to anaerobic, respiration; aerobic respiration produces many more energy-rich molecules per unit of glucose oxidized than anaerobic respiration. As a result, organisms whose metabolism is based on aerobic respiration can grow faster and live more active lives than anaerobes. Whereas life on Earth had consisted of many kinds of single-celled microorganisms for about 3.6 billion years, all of a sudden (relatively speaking), as the fossil record reveals, the oceans of the world became populated by a diverse array of new multicellular forms of life, including early members of familiar animal phyla such as annelids, mollusks, and arthropods that continue to exist today. Also among these were fishlike chordates, the phylum to which the vertebrates belong. So our vertebrate heritage extends back at least about 500 million years in marine ecosystems.

ON TO THE VERTEBRATES

Along with many plants, vertebrates are the largest terrestrial and marine organisms on Earth today and have been so for about four hundred million years. Fish, particularly jawless forms, were the first vertebrates. Amphibians were the first tetrapods and first terrestrial vertebrates, appearing by about 350 Ma (million years ago). Early on they likely coexisted with early reptiles, which traditionally have included the synapsids that gave rise to mammals and true reptiles, about 360–350 Ma (in the Early Carboniferous period). Synapsids are now considered to be the sister group (i.e., apart from but sharing a common ancestor) of reptiles. True reptiles, which currently include crocodiles, turtles, snakes, and lizards, underwent most of their extensive and remarkable evolutionary radiation during the Mesozoic era (252–66 Ma; table 2). Fur-bearing mammals evolved from advanced synapsids known as therapsids in the Early Jurassic period about 200 Ma, and birds evolved from archosaurian reptiles (and are actually feathered dinosaurs) in the Late Jurassic about 150 Ma. For most of the Cretaceous period (i.e., from 145 to 66 Ma; table 2), however, early mammals and birds were overshadowed in size and diversity by dinosaurs and their relatives. It was only after the asteroid impact that devastated all of Earth's ecosystems at the boundary between the Cretaceous and Tertiary periods (at 66 Ma)—the so-called K/Pg boundary—that birds and mammals of increasingly modern aspect became the dominant terrestrial vertebrates. While the asteroid impact had a profoundly negative effect on much of life on Earth, as had four previous mass extinction events, in hindsight it ultimately had a positive effect by clearing life's slate for the rise of modern birds and mammals, including us.

Finally, how did the transition from an aquatic existence to a terrestrial existence take place? To start with, the fossil record provides us with a fairly detailed picture of the morphological changes that occurred during this transition. We thus know that the bony lobe fins of certain predatory marine and freshwater fish—the sarcopterygians (also known as crossopterygians), which are represented today by freshwater lungfish and marine coelacanths—developed into tetrapod limbs gradually over a period of some twenty million years. Their pectoral fins supported by a pectoral girdle were first to develop as limbs; robust pelvic limbs developed later. Most discussions of this transition suggest that these transitional forms lived in shallow bodies of water, most likely freshwater, and that their front limbs initially allowed them to occasionally forage along the edges of shallow lakes and swamps. Access to new food sources in addition to avoidance of competition with and predation by other fish were some of the factors that may have selected for this transition. As you can imagine, many profound morphological and physiological changes had to take place during this transition from life in a buoyant watery medium in which gills were used to obtain oxygen to life in a non-buoyant, gravity-driven medium in which lungs were used to obtain oxygen.

GREENING OF THE EARTH

Prior to the evolution of terrestrial multicellular life, only single-celled microorganisms lived on land. Terrestrial microbial mats are known from about 2.6 Ga. So from at least this point in time until about 0.48 Ga, we must visualize a barren landscape on Earth devoid of terrestrial fungi, plants, and animals. Nonetheless, green algae, the eukaryotic ancestors of land plants, were present in oceans by about 725 Ma. According to the fossil record, the first terrestrial plants predated the first terrestrial vertebrates and appeared about 475 million years ago. These plants were small, lacked a water-transporting vascular system, and probably resembled today's mosses and liverworts; they reproduced via spores. Over the next one hundred million years, they began to evolve larger, treelike kinds and to form forests that were also inhabited by Earth's first insects and other arthropods. These animals, in turn, became important prey items for early terrestrial vertebrates. The greening of Earth's terrestrial surfaces occurred for the first time during this interval.

Before the evolution of angiosperms (flowering plants) at about 150–130 Ma in the Late Jurassic and Early Cretaceous periods (table 2), Earth's terrestrial plant flora underwent an extensive radiation involving the evolution of vascular systems, leaves with stomata, and roots. Vast coal-forming swamps inhabited by giant spore-producing lycophytes (relatives of club mosses) and horsetails as well as ferns and extinct taxa such as pteridosperms, sphenopsids, and glossopterids were prominent plants of the Carboniferous period (359–299 Ma). After that, forests containing various lineages of seed-producing gymnosperms covered both wet and dry environments throughout the Permian, Triassic, and Jurassic periods—a timespan of about 150 million years.

Charles Darwin (1809–82) called the sudden appearance of angiosperms in the fossil record an "abominable mystery" primarily because, taking this record at face value, it conflicted with his view that all of life evolved in slow, gradual fashion. Nonetheless, clear evidence that this new group of plants, which bore their seeds in protected carpels rather than as naked seeds in cones, had evolved by the Early Cretaceous is overwhelming. Where on Earth they first evolved and the steps involved in the transition from naked seeds to protected seeds produced in flowers are still being investigated. But in the end, the occurrence of flowers and their seed-containing fruits proved to be a huge evolutionary success. By the Late Cretaceous, angiosperms were well on their way to becoming the dominant land plants on Earth. As a result, Earth's plant communities today contain over three hundred thousand species of angiosperms compared with only about one thousand species of gymnosperms and over twenty thousand species of nonvascular plants. One important reason for the evolutionary success of angiosperms comes from their frequent reliance on animals to disperse their pollen grains and seeds. Animal dispersal of gametes and propagules generally gives angiosperms much greater mobility than plants that rely on gravity, wind, or water to disperse their gametes, seeds, or spores.

2

TREE THINKING

THE BASICS

THE IDEA THAT THE history of life on Earth can be represented by a treelike metaphor has been a part of our biological psyche at least since the time of Charles Darwin, who is well known for producing a treelike sketch of how he envisioned evolution to proceed in one of his notebooks in 1839. Since then, a major goal of evolutionary biology has been to produce an ever-more-detailed picture of how current and past organisms are related to each other and how life has unfolded on Earth. The final goal of this pursuit would then be the creation of a "tree of life." However, two major discoveries toward the end of the twentieth century have raised serious questions about the validity of the tree metaphor. As I've mentioned, the first discovery involves horizontal gene transfer (HGT) that has been occurring among unrelated organisms (e.g., between archaea and bacteria) throughout the history of life. The second one is that eukaryotes are mosaics of endosymbiotic organelles and their genomes as well as some genetic material from retroviruses. As David Quammen suggests in *The Tangled Tree,* perhaps life's history is best portrayed as a web of interconnections rather than simply as a tree.

Despite our current uncertainty about how to best portray the history of life, modern biology has been revolutionized by a more modest form of "tree thinking" in which relationships between a relatively limited group of organisms and their adaptations are summarized in a treelike phylogeny representing a hypothesis of the evolutionary history of that group and its traits. This approach to the study of evolutionary relationships and the evolution of adaptations of all kinds has occurred as a result of the confluence of large amounts of data from DNA and morphological studies as well as from the fossil record. Combining these datasets using sophisticated analytical techniques has produced a vast literature on the detailed and time-calibrated evolutionary histories of various groups that would have astounded Darwin if he were alive today. To set the stage for detailed discussions of the evolution and coevolution of nectar-feeding birds and mammals and their food plants, I will next present brief summaries of their broad evolutionary histories based on their current phylogenies.

THE EVOLUTIONARY HISTORY OF FLOWERING PLANTS

Resolving the relationships between taxonomic orders and families within the flowering plants has long been a goal of systematic botanists. Early attempts, of course, had to rely mostly on morphological traits of fossils and living material, but that has changed dramatically with the advent of molecular (mostly DNA) phylogenetics. When large numbers of genes or entire genomes are used for determining evolutionary relationships, large-scale analyses of evolutionary relationships are possible. Thus, beginning in the 1990s an international group of systematic botanists—the Angiosperm Phylogeny Group (APG)—began to create a classification of flowering plants based primarily on molecular genetics techniques. To date, four iterations of this classification (APG I–IV) have been published (in 1998, 2003, 2009, and 2016). My botanical colleague, John Kress, and I used APG III as the basis for our comprehensive study of tropical vertebrate/plant coevolutionary interactions in *The Ornaments of Life* published in 2013. I will use this classification and the results of many of our analyses throughout this book. According to Douglas Soltis and colleagues, APG IV is not much different from APG III, so the use of the earlier version seems justified.

A phylogenetic (evolutionary) tree depicting the relationships between taxonomic orders of angiosperms is shown in figure 1. In this classification, five groups of orders were identified. Three of them—monocots (e.g., bananas and bromeliads), asterids (e.g., primroses and daisies), and rosids (e.g., roses and legumes)—are monophyletic, meaning that all members of their orders share a common ancestor. The other two groups—basal angiosperms (e.g., water lilies and magnolias) and basal eudicots (e.g., cacti and broccoli)—are not monophyletic, meaning that their orders do not share a common ancestor. They are grouped together because they share certain "primitive" (for that group) traits or characteristics. APG does not attempt to put a timescale on their phylogenies, but the relative lengths of the branches in figure 1 give us a picture of the relative ages of the groups and their orders. Branches associated with the basal angiosperms are obviously much longer than those associated with the rosids, implying that orders (and their families) in the former group are much older than those in the latter.

In 2015, Susan Magallón and colleagues published a time-calibrated estimate of the ages of about 87 percent of angiosperm families using several DNA markers from chloroplasts and the nucleus. An extensive fossil database was also used for the time calibration. Results indicated that many angiosperm families began to diversify in the Early Cretaceous (well before the asteroid impact) (table 2). Basal angiosperms, monocots, and basal eudicots of APG III began to diversify 135–130 million years ago; rosid diversification began 123–115 million years ago; and asterids began diversifying 119–110 million years ago. These estimates suggest that many of the morphological and functional traits of these groups—traits that often define different families—are old, although their current species are likely to be much younger.

If this time calibration is correct, what does this tell us about the speed with which the diversification of families and orders of flowering plants has occurred? In our mind's eye, I think we can visualize the vast differences in, for example, leaf and flower morphology between magnolias, with their broad, shiny, and simple leaves and large showy flowers, and legumes, with their finely dissected pinnate leaves and shaving brush or pealike flowers. These are the kinds of characters or traits that have traditionally been used to define different groups of angiosperms. Fossil evidence suggests that early angiosperms were small shrubby or herbaceous plants with small (1–2 mm), radially symmetrical flowers (picture a sunflower) and small seeds. Early on their flowers were visited and likely pollinated by flies, beetles, and other early insects searching for a bit of nectar and pollen to eat. Today, most flowering plants are still insect pollinated, and bees, ants, and wasps and their relatives (order Hymenoptera), butterflies and moths (order Lepidoptera), and various kinds of flies (order Diptera) are their major pollinators. Although these three groups probably were not the earliest pollinators of angiosperms,

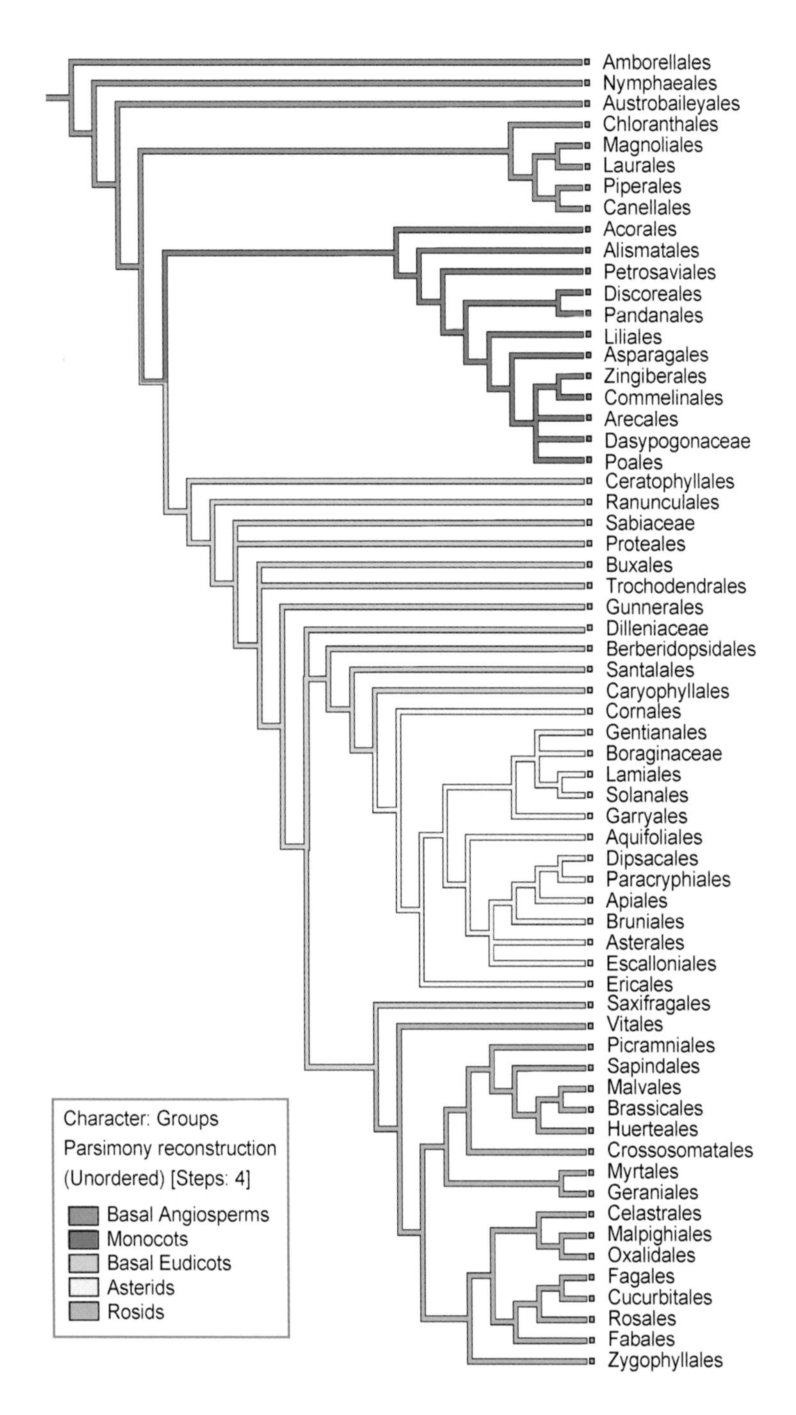

FIGURE 1 The phylogenetic history of flowering plants based on APG III (2009) in which five major groups are highlighted. From Fleming and Kress (2013) with permission.

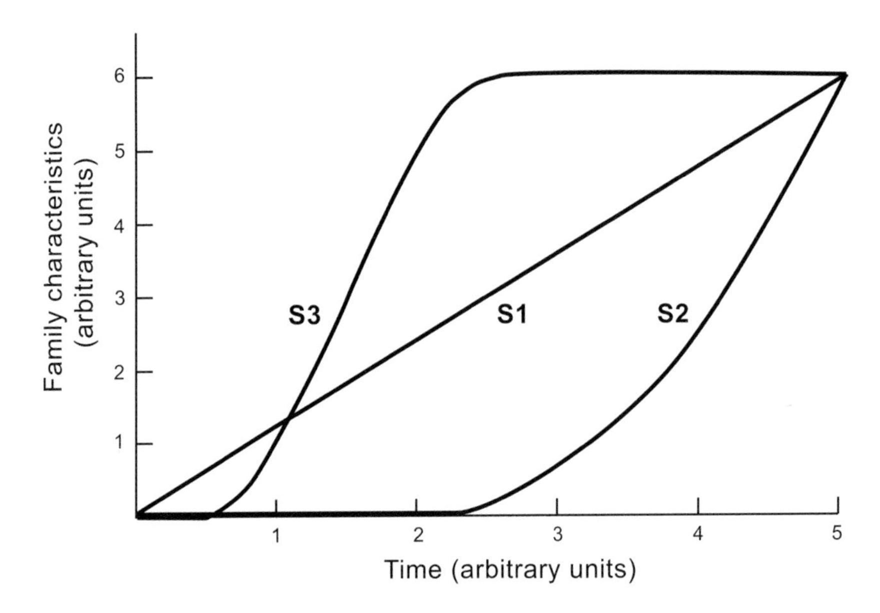

FIGURE 2 Three hypothetical evolutionary scenarios showing the acquisition of a plant family's defining morphological characteristics. These kinds of scenarios can also be applied to the acquisition of a taxonomic group's overall species diversity, among many other biological topics. Courtesy of Steve Dyer.

fossils tell us that they were present during the early evolution of these plants. Recent phylogenomic research (i.e., phylogenies built using genomic data), for example, dates the orders containing bees, butterflies, and flies as being about 250, 150, and 150 million years old, respectively. The presence of these potential pollinators undoubtedly was important for floral evolution during the initial radiation of flowering plants in the Early and Late Cretaceous (table 2).

So what might this radiation have looked like in terms its speed and shape? It's likely that each family has had its own evolutionary trajectory, but it's easy to envision at least three hypothetical scenarios, as illustrated in figure 2. Please note that the time axis (X) in figure 2 is not precisely defined; it could be in thousands or millions of years. But the shape of each scenario is important. In S1, the trajectory is simply a steady increase through time toward a family's final morphological diversity or to a point where a family's defining characteristics have been attained. S2 depicts a long period of stasis followed by the rapid attainment of a family's defining characteristics. And

S3 is the opposite, that is, the early and rapid attainment of a family's defining characteristics. Variations on these scenarios probably are myriad, and it is likely that different parts of plants (e.g., flowers, leaves, growth habit, etc.) have had their own temporal trajectories that reflect the intensity of different selection pressures and the strength of genetic-developmental constraints. For example, flowers and their characteristics have been much more likely to respond quickly to biotic selection pressures than are the stem and root morphology of angiosperms.

These kinds of questions about the tempo and mode of evolutionary changes are general to all of life, and we will see them again in specific discussions of the evolution of nectar-feeding birds and bats and their food plants. The general scenarios illustrated in figure 2 have also been used in discussions of the diversification of particular groups of organisms (e.g., birds or mammals). S2 has been called a "long-fused" model of diversification, and S3 has been called a "short-fused" model. We will revisit this topic below.

THE EVOLUTIONARY HISTORY OF BIRDS

Birds are the only surviving lineage of the once dominant nonavian dinosaurs. Consequently, they are now considered to be "feathered dinosaurs." This realization is very recent and has emerged from the discovery of the fossils of many kinds of true feathered dinosaurs in northeastern China and elsewhere. Based on these fossils, paleontologists have been able to trace the development of many of the morphological traits that define birds in their theropod dinosaur ancestors over a period of about one hundred million years. These traits include the loss of a fifth toe on the hind foot, the evolution of the furcula (a modified clavicle), the presence of feathers with vanes, and a keeled sternum. At face value, this trajectory appears to resemble scenario one in figure 2. That is, the acquisition of avian characteristics likely took place gradually over a long period of time. They didn't emerge quickly during the radiation of feathered dinosaurs. All of these changes make me wonder about the selective pressures behind them. For example, what biotic or abiotic conditions were crucial to the evolution of feathers, the single most definitive characteristic of birds? Were they originally valued for their insulative properties, for their social signaling value, or what? They certainly didn't initially evolve so that dinosaurs could fly.

Finally, at an age of about 150 Ma in the Late Jurassic (table 2), the well-known fossil *Archaeopteryx lithographica* was discovered in the fine-grained shales of a lagoon in tropical Europe. This crow-sized species is still considered to be a reptile but has many of the features found in the very earliest true birds, including fully modern feathers in its wings and long tail. However, its lack of an avian shoulder joint and a keeled sternum for the attachment of large flight muscles indicates that at best it was a weak flier and was incapable of strong, sustained flight involving rapid wing beats. A bit later in the Cretaceous, true birds gained their full suite of avian morphological characteristics incrementally as they evolved into a substantial variety of ecological and behavioral types. These birds ranged in size from sparrows to vultures. Ecologically, they included wading birds, diving birds, arboreal perching birds similar to our modern songbirds, and a few flightless kinds. Along with the dinosaurs and many other kinds of vertebrates, however, nearly all of these early birds were doomed to perish as a result of the asteroid impact.

After the mass extinction of many forms of life at 66 Ma, the ecological "slate" of many/most of Earth's ecosystems was wiped nearly clean, as had also occurred in previous mass extinctions (e.g., the Permian-Triassic mass extinction that occurred at 252 Ma; table 2). Nonetheless, this devastating event presented tremendous new ecological opportunities for the survivors, and many groups, including angiosperms, birds, and mammals, certainly took advantage of this, just as early dinosaurs did after the Permo-Triassic extinctions.

Modern birds are classified in three monophyletic groups: Palaeognathae (tinamous, flightless ostriches, and their relatives), Galloanserae (waterfowl, chickens, and game birds), and Neoaves (all other birds). Of these, paleognaths first evolved in the Late Cretaceous before the impact (at about 82 Ma) whereas the other two groups originated just after the impact (after 66 Ma). Diversification of the many orders of Neoaves was rapid (short-fused; figure 2) and by about 47 Ma, the large order Passeriformes (the songbirds), which contains about 60 percent of current bird species, had begun to radiate in Australasia. The evolution of the diversity of Neoaves from grebes and flamingos at its base to the beginning of the passerine radiation at its apex took place in only about twenty million years and is a truly amazing demonstration of how quickly life on Earth can fill an ecological vacuum. We'll explore some of the mechanisms involved in these evolutionary changes a bit later.

THE EVOLUTIONARY HISTORY OF MAMMALS

Like birds, mammals have had a long evolutionary history dating back some three hundred million years (to the early Permian) with the rise of synapsids, early terrestrial tetrapods that are the sister group of reptiles. Synapsids dominated terrestrial faunas worldwide during the Permian and early Triassic periods, but many went extinct at the end of the Permian (table 2). During their heyday, synapsids included both herbivores (e.g., the pelycosaur *Cotylorhynchus*) and carnivores (e.g., the pelycosaur *Dimetrodon*). These were basically large animals, reaching a maximum length of about three meters. One group of advanced synapsids, the therapsids, survived the Permo-Triassic extinction but gradually decreased in importance toward the end of the Triassic. Before it disappeared, however, therapsids gave rise to a group called cynodonts (from the Late Triassic at about 205 Ma).

Beginning with the therapsids, synapsids began to exhibit some basic mammalian skeletal traits, including heterodont dentition (i.e., with teeth differentiated into incisors, canines, and molar-like teeth), many advanced skull and braincase features, and in some, a change from a sprawling limb posture to a more upright pillar-like limb position. Further modifications of the synapsid skeleton, including a new articulation between the lower jaw and skull, development of a secondary palate in the skull, evolution of the middle ear, a reduced number of ribs, and a larger brain (and by implication, changes in their physiological, digestive, and cognitive features), occurred in cynodonts, which are the direct ancestors of mammals that appeared in the Late Triassic and Early Jurassic (table 2). Unlike their early synapsid ancestors, Mesozoic mammals were mouse- to occasionally cat-sized. They clearly lived in the shadow of dinosaurs and were nocturnal quadrupedal insectivores or rodent-like seed eaters (e.g., the multituberculates). They were now well furred, presumably endothermic, and gave birth to live young that they nursed from mammary glands.

As in birds, most modern mammals radiated in the postimpact Cenozoic era. But the three well-known modern mammalian lineages—monotremes, marsupials, and placentals–clearly began to evolve well before the impact in the Cretaceous. Marsupials and placentals are sister groups that have had very obvious differences in their evolutionary success. Whereas the approximately 334 extant species of marsupials (in 19 families) currently include mostly mouse- to kangaroo-sized terrestrial or arboreal species (although much larger kinds used to live in Australia), the size range of contemporary placentals with about 4,000 species in about 114 families is enormous and includes many aquatic and aerial kinds in addition to terrestrial and arboreal species. Results of a recent morphological/DNA-based phylogenetic analysis indicates that modern forms of marsupials and placentals began to radiate in the early Paleocene, shortly (2–3 million years) after the impact. Most current orders of placentals had appeared by 55–50 million years ago.

In summary, the postimpact adaptive radiation of both modern birds and mammals, like that of angiosperms, occurred relatively rapidly in geological time. These radiations therefore more closely resemble a "short-fuse" scenario (S3) than the other scenarios depicted in figure 2. As a result, these groups quickly filled available ecological space. Major driving factors behind this undoubtedly involved some combination of competition for limited resources, predation, and mutualistic interactions with flowering plants in addition to the significant climatic changes that have occurred during the Cenozoic era. I will explore the consequences of these factors as they apply to nectar-feeding birds and bats and their food plants in the rest of this book.

3

HOW TO BUILD A HUMMINGBIRD

AN OVERVIEW OF HUMMINGBIRD DIVERSITY AND EVOLUTION

ALONG WITH NIGHTJARS (INCLUDING our familiar nighthawks), frogmouths, potoos, oilbirds, and owlet-nightjars, swifts (family Apodidae, 96 species) and hummingbirds (family Trochilidae, 368 species) have traditionally been included in the neoavian order Caprimulgiformes. But recent phylogenetic analyses have placed the Caprimulgiformes and several other related orders into a more broadly construed clade (i.e., a group of taxa that share a common ancestor) called Strisores (figures 3 and 4). Swifts are the sister group to hummingbirds, and these families have traditionally been classified together in the order Apodiformes, one of the orders in Strisores. This is a very interesting evolutionary grouping because most members of the Strisores are nocturnal and mostly insectivorous (except the fruit-eating oilbird of northern South America) whereas swifts and hummingbirds are diurnal insectivores (and nectar eaters, of course, in hummingbirds). Despite differences in their activity times, most of these birds are characterized as having long pointed wings, small feet, and short weak bills (except for hummingbirds). None of them have much ability to walk on the ground. The nocturnal forms are cryptically colored and have large eyes, soft, lax feathers, and broad but weak bills (except for oilbirds, which are strong billed) for scooping up their insect prey out of the air at night. You've probably seen nighthawks chasing moths around streetlights and lighted billboards, sometimes in the company of insectivorous bats. Nightjars have a worldwide distribution; swifts occur in all continents except for Australia; and hummingbirds reside only in the New World.

Although hummingbirds now occur only in the New World, their scanty fossil record indicates that they once also occurred in tropical Europe in the Early Oligocene, some 32–28 million years ago. Finding these fossils in Germany was so surprising that the paleontologist Gerald Mayr in 2004 named the first specimen *Eurotrochilus inexpectatus*. This beautifully preserved fossil had the same specialized wing, shoulder, and bill morphology as modern hummingbirds, which strongly suggests that they were also able to hover and visit long-corolla flowers for nectar. The implication here is that bird-plant coevolution was already well underway in tropical Europe ten million years before hummingbirds began to evolve in the New World. *Eurotrochilus* differed from similar-aged swifts by its very small size and its long bill—features that still characterize modern hummers.

The hummingbird and swift families separated about forty-two million years ago (in the Eocene; table 2), and this split most likely took place in Eurasia, where swift diversity is currently highest. A recent phylogenetic analysis by Jimmy McGuire and colleagues indicates that the new-world radiation of hummingbirds began about twenty-two million years ago in the lowlands of South America. This obviously begs the question, when and how did these birds get from Eurasia to South America? McGuire and colleagues suggested that the most likely route was via Beringia and then down through North and Central America before island-hopping into South America. How long this took and what kinds of flowers they fed at along the way are still unanswered questions. Since hummingbirds today feed at flowers that are morphologically specialized, we're left to wonder what kinds of flowers they might have fed at while traveling south through the Americas. They certainly didn't encounter these kinds of flowers when they first entered North America because insects were the main pollinators of flowers then as they still are today. Since many hummingbirds today, especially those with short bills, regularly visit insect-pollinated flowers, it is likely that these kinds of flowers were the original energy sources for hummingbirds initially entering North America.

According to the McGuire and colleagues age-dated phylogeny (figure 3), once in South America, hummingbirds diversified into nine distinct clades or subfamilies. The basal (or oldest) clades—Topazes and Hermits containing a total of about 40 species—are mostly South and Central American in

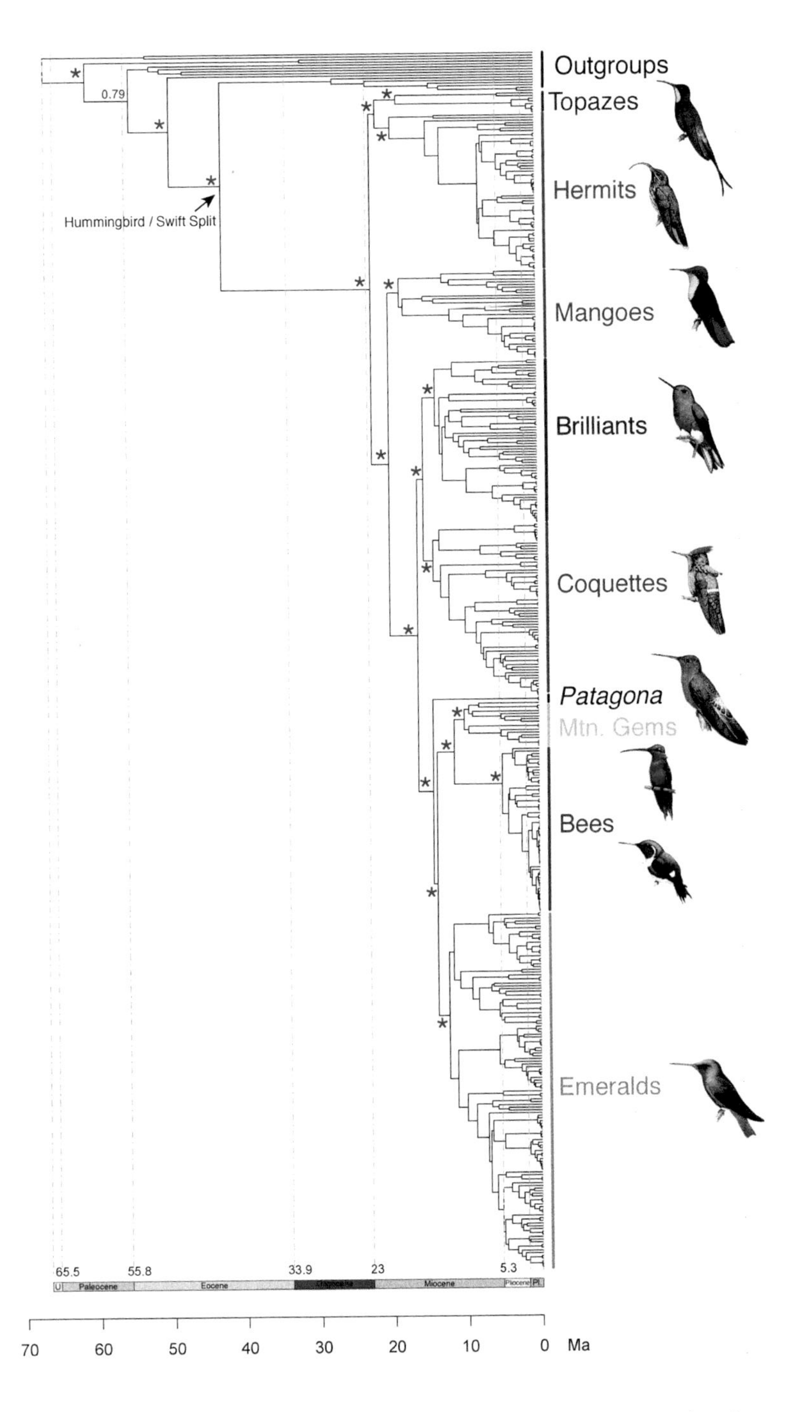

FIGURE 3 A dated evolutionary history of hummingbirds. From McGuire et al. (2014) with permission.

distribution. In contrast, the most advanced (youngest) clade—Emeralds containing about 108 species and originating about 12.5 million years ago—contains species that are found in South, Central, and North America as well as the Caribbean. A notable event in the radiation of hummingbirds was the colonization of the northern Andes as they began their final uplift 10–12 million years ago. About 140 species, primarily in three clades (a single species of *Patagona* plus the Coquettes [61 species] and Brilliants [51 species]) are restricted primarily to the Andes (figure 3). Mangoes colonized the Caribbean about five million years ago, and ancestors of the Bees and Mountain Gems began to colonize North America about 12 million years ago.

An important reason why the hummingbird family is one of the largest bird families today is the size of their geographic ranges. Species of hummingbirds generally have much smaller geographic ranges than other birds. This is especially true of Andean species, many of which occur in single valleys or on particular mountaintops. These beautiful and highly specialized birds seem to have filled nearly all of their available ecological niche space and are even able to persist in small habitats, especially in the Andes. One wonders how the Andean forms will be affected by climate change and global warming, which tend to have particularly strong negative effects on montane and high-latitude habitats.

Finally, what does a curve depicting the accumulation of species in the nine hummingbird clades through time look like? Which of the three curves in figure 2 best shows this? It turns out that scenario S1, a steady accumulation of species through time, best depicts the data from this phylogeny. This diversification occurred at a rate of about 0.25 species per million years or 1 new species every four million years. But a closer look at the diversification rates within specific clades indicates that the relatively young Bee clade (figure 3) has had a notably higher rate of diversification than much older clades such as Topazes, Hermits, and Mangoes. Why this is the case is currently unknown.

THE EMERGENCE OF SWIFTS AND HUMMINGBIRDS

As I've mentioned, hummingbirds and swifts became separate avian families beginning about forty-two million years ago. This evolutionary split likely occurred in Eurasia at a time when it was generally covered in tropical moist forest at lower paleolatitudes and drier habitats at higher paleolatitudes.

Occupying a large portion of the former northern supercontinent of Laurasia, Eurasia likely also had important dispersal connections with parts of the former southern supercontinent of Gondwana, especially Africa and India, early in the Cenozoic. As a result, it was an important evolutionary theater for the evolution of many kinds of plants and animals, including bats and cats as well as many kinds of reptiles. Overall, this suggests that Eurasia was especially rich in ecological opportunities that promoted the adaptive radiation of many modern forms of life on Earth.

Phylogenetic analyses plus the fossil record have provided us with a reasonably complete picture of the early stages in the evolutionary split between swifts and hummingbirds as well as the rest of the Strisores (figure 4). Luckily, early nightjars, swifts, and their relatives are relatively well-represented in the fossil record, especially in Europe. This is surprising because many of them were rather small and delicate (especially early hummingbirds, of course). So

FIGURE 4 Four examples of Strisores: A. lesser nighthawk (*Chordeiles acutipennis*), B. Australian owlet-nightjar (*Aegotheles cristatus*), C. common swift (*Apus apus*), and D. Talamanca hummingbird (*Eugenes spectabilis*). Photo credits: Dan Weisz (A); Adobe stock (B, C); Ted Fleming (D).

they might be expected to be poorly represented by fossils among early neo-avian birds. Nonetheless, some of them have turned up in fine-grained shales associated with freshwater habitats. As a result, many of their fossils have been preserved in exquisite morphological detail, including their feathers in some cases. I find published images of these birds, their delicate bones set in a stony matrix, to be a thrilling look back into the distant past and a strong affirmation of the subtle pageantry of the history of life on Earth.

Two of these fossils include *Parargornis messelensis* from the Early Eocene (at 49 Ma) in Germany and *Eocypselus rowei*, from the Early Eocene (52 Ma) in Wyoming. Both of these taxa are considered to be basal (ancestral) to the actual divergence between swifts and hummingbirds. As a result, they give us important insights into what the ancestors of these two kinds of birds were like. *Parargornis messelensis* was a small bird with a short, wide beak, short

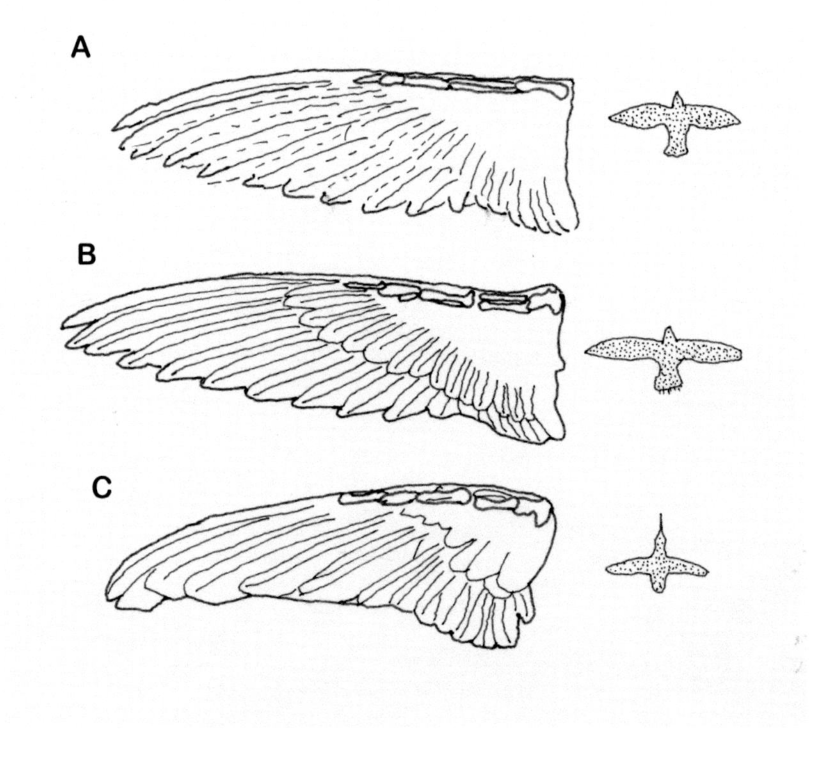

FIGURE 5 Diagrams of the wings of three Strisores: A. the fossil *Eocypselus rowei*, an ancestor of swifts and hummingbirds, B. a modern swift *Hirundapus caudacutus*, and C. a modern hummingbird *Archilochus colibri*. Redrawn from Ksepka et al. (2013).

rounded wings, and long legs with long toes. Its wing skeleton featured a very short humerus and an elongated set of hand bones, as occurs in modern swifts and hummingbirds. It probably could perch on branches and sally out after insects and walk on the ground. *Eocypselus rowei* was also a small bird (the size of a modern chimney swift weighing about 23 g) with a short, rounded beak and wing morphology that was intermediate between the long pointed wings that swifts use for fast-flapping and gliding flight and the shorter, rounder wings that hummingbirds use for hovering (figure 5). Like *P. messelensis, E. rowei* had long legs and toes for perching and possibly terrestrial locomotion.

Using an extensive molecular and fossil database, Chen and colleagues in 2019 summarized the phylogenetic history of the Strisores (formerly Caprimulgiformes and containing seven families, including swifts and hummingbirds) (figure 6). Results of their analyses indicate that aerial foraging and a nocturnal lifestyle are ancestral in this group and that diversification into different lineages likely occurred rapidly in the Eocene. As in other recent phylogenetic

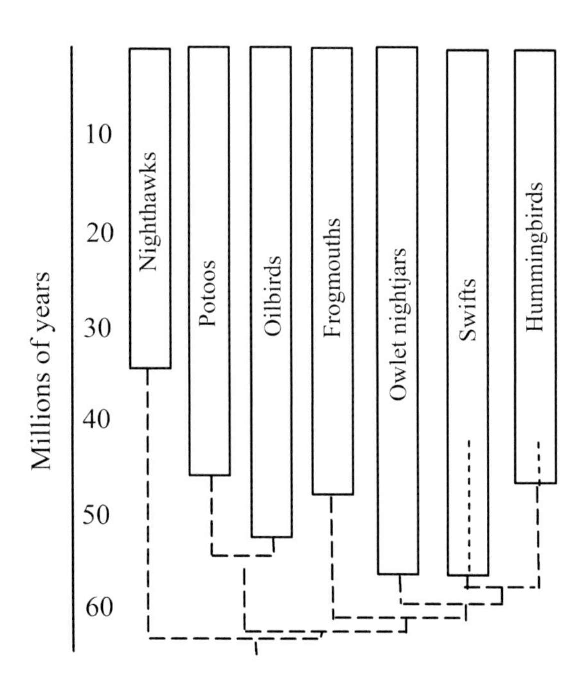

FIGURE 6 A dated evolutionary history of the avian order Strisores, which includes nightbirds, swifts, and hummingbirds. The small, dotted lines in the swift and hummingbird boxes indicate the presence of ancestors of those two families. Redrawn from Chen et al. (2019).

analyses of this group, owlet-nightjars (Aegothelidae), currently found only in Australia and New Guinea, are the sister group of the Apodiformes (swifts and hummingbirds). These three families last shared a common ancestor in the Paleocene about 63 million years ago (figure 6).

Studies that detail the evolutionary history of a group of organisms often make me wonder about how these radiations unfolded. What were the selective pressures that caused different lineages (and ultimately, different but related families) to evolve, and what were the genetic-developmental processes that produced the morphological differences that we see in them today? In terms of the gross features of these radiations, it is easy to speculate that they occurred by one or more species moving into and eventually filling "empty" ecological space with its descendants. Most birds today are active during the daytime, and relatively few kinds are nocturnal. The two prominent groups of nocturnal birds include owls, which are basically sit-and-wait vertebrate predators; they are the nocturnal counterparts of diurnal falcons and hawks. The other group includes nightjars and their allies, which are aerial pursuers of flying insects (figure 4). Both of these groups began to evolve in the early Cenozoic (about 57 to 60 million years ago), and they apparently were quick to fill the ecological space available to nocturnal avian predators. It wasn't until much later, in the Miocene (table 2), that swallows, a passerine family of diurnal aerial insectivores, began to evolve. Today, like swifts, swallows (family Hirundinidae with twenty genera and eighty-six species) have a worldwide distribution.

But what were the evolutionary transitions that produced, for example, different kinds of nocturnal aerial hawkers and eventually, diurnal hawkers (swifts) and nectar feeders (hummingbirds)? What are some of the genetic-developmental mechanisms behind these transitions? How quickly did these transitions take place? Although he knew nothing about genes and developmental pathways, Charles Darwin knew from the work of pigeon fanciers (and his own captive birds) and horticulturalists that selective breeding often quickly produces substantial and dramatic morphological changes in these organisms. And from this he surmised that given enough time, natural selection could also produce these kinds of morphological changes in nondomestic organisms. Genetic variation and biotic and abiotic environmental selective pressures are what is needed to effect these changes.

Our current understanding of the genetic-developmental events behind the evolution of complex adaptations is constantly growing owing to all of the advances that have occurred in the fields of molecular genetics and development in recent decades. When placed in a phylogenetic context, this information can provide us with a fascinating view of how the evolution of adaptations actually occurs. But rather than discuss these scientific advances in detail (they can be found in modern evolutionary textbooks, for example), suffice it to say that we know that the genetic bases of adaptations are complex and involve gene duplications and development of new functions in these duplications. They also involve intricate interactions among many genes and their regulatory and enhancer elements with particular genes being expressed only under certain conditions in a developmental pathway.

As an example, let's consider the complex genetic-developmental pathways found in bilateral (rather than radial) multicellular animals such as arthropods and tetrapods. These animals last shared a common ancestor at least five hundred million years ago. So we might not expect them to have similar genetic pathways in their body form evolution. But we would be wrong. It turns out that in both groups the genes that control their anterior to posterior development are dictated by HOX genes, which are a series of genes that specify the development of particular regions of the body plan of embryos along the head-tail axis. They are responsible for creating segmentation in invertebrates and the placement and development of limbs and other appendages in arthropods and tetrapods. They do this by controlling cascades of other interacting genes and tend to be highly conserved but malleable over evolutionary time. Through time, for instance, duplication and favorable mutations in these genes and the genes they control have led to the evolution of tetrapod limbs from the fins of sarcopterygian fish. So in a sense, evolution at the genetic-developmental level tends to be conservative. New adaptations arise from the reworking of older ones whose genetic-developmental pathways have changed owing to the occurrence of favorable new mutations, new genetic interactions, and new selective pressures.

Given this background, let's now consider the evolution of feeding-related adaptations seen in some well-studied birds, including the Strisores. An obvious place to begin these studies would be in geologically young islands where well-known evolutionary radiations in their birds have occurred over short periods of time. These choices would include the Hawaiian and Galapagos Islands, which are only about 5–7 million years old or less. Both of these young archipelagos contain (or contained) an impressive morphological diversity of closely related birds. In a period of about 7.2 million years, for example, the Hawaiian honeycreepers (Fringillidae, Carduelinae) evolved into about fifty-seven species classified in twenty-four genera, of which about twenty-one species are now extinct owing to the hand of man. This radiation produced a variety of feeding types, including thick-billed seed eaters (the ancestral type), thin-billed insectivores, and curve-billed nectar feeders. In the Galapagos, thirteen species in four genera of Darwin's finches (Thraupidae) evolved in about three million years or less. This radiation also produced a variety of feeding types, including insectivores, cactus flower visitors, and small and large thick-billed seed eaters (but no morphologically specialized nectar feeders). Recent detailed observations, however, indicate that many of these birds, including seed eaters, regularly visit and presumably pollinate flowers.

To begin to understand the impressive radiation of feeding types among Hawaiian honeycreepers, Irby Lovette and colleagues published a study in 2002 comparing the degree of genetic and morphological variation that exists in these birds and their mainland ancestors with that of Hawaiian thrushes (Turdidae) and their relatives. In contrast to Hawaiian honeycreepers, the evolution of Hawaiian thrushes has produced only five very similar species in a single genus in about the same amount of time. And they all still look like our familiar backyard robins. Their results showed that Hawaiian honeycreepers are derived from a group of birds that exhibit significantly more morphological (and presumably genetic and developmental) variation in their bills than thrushes, an evolutionarily older family of birds. The suggestion here is that, as a group, both mainland and island cardueline finches have had a greater ability to evolve novel bill and feeding adaptations than thrushes. Their bills appear to be more evolutionarily labile than those of thrushes. Hence, this inherent plasticity allowed Hawaiian honeycreepers to evolve into a greater variety of feeding niches than Hawaiian thrushes once they began to colonize these new, ecologically depauperate islands. Still unanswered from this early study, however, is what is the genetic-developmental basis for this plasticity?

To explore this topic in more detail as it relates to island honeycreepers and finches and the ancestrally nocturnal Stisores, including swifts and hummingbirds, let's review the results of two recent morphological studies of these birds conducted by Guillermo Navalón and his colleagues. In 2020, this group sought to determine whether adaptive morphological changes in the skulls of birds, especially those of the quickly evolving Hawaiian honeycreepers and Darwin's finches, involves tight integration between beak and skull morphology (scenario one) or low integration between these two structures (scenario two). If low integration exists, this implies that different parts of the skull (i.e., different morphological modules, including beaks and braincases) are free to evolve independently of each other. A priori, it might be expected that skull morphological evolution in birds should proceed more rapidly under scenario two than under scenario one because of the apparent of lack of genetic-developmental constraints implied in scenario two. Although the actual genetic-developmental basis of skull evolution was not examined in this study (as it has been in the evolution of beaks of Darwin's finches), the existence of these two scenarios clearly has important genetic implications. Scenario one implies that a single genetic-developmental program likely underlies skull evolution whereas scenario two implies that different parts of the skull (i.e., different modules) are controlled by somewhat different genetic programs.

Results of this study, which examined the degree of morphological integration within skulls of representatives of all modern bird lineages, indicated that morphological integration was much lower (i.e., scenario two) in nonpasserines than in passerines. As a result, diversity of skull shapes (and the range of feeding adaptations) is much greater in nonpasserines than in passerines. Despite the great taxonomic diversity of passerines, they show very little diversity in skull shapes with the notable exceptions of Hawaiian honeycreepers and Darwin's finches. Different parts (modules) of the skulls of passerine birds appear to be highly integrated (scenario one), and species in the two islands exhibit exceptionally high levels, not low levels, of morphological integration. Thus, contrary to expectations, rapid adaptive evolution in the skulls of these birds appears to conform to scenario one: natural selection is pushing skull evolution along "lines of least resistance" (to quote Navalón et al.) toward new adaptive peaks. Different modules of skull morphology are not free to vary independently of each other. And by implication, avian skull development in these birds (and in passerines generally) appears to be controlled by a single genetic program rather than by several separate programs.

The study published by Navalón and colleagues in 2021 dealt directly with craniofacial integration in the skulls of birds in the Strisores. It examined in great detail the ontogenetic (developmental) changes in the skulls of representatives of the three monophyletic groups of modern birds (Palaeognathae, Galloanserae, and Neoaves; see page 18) with special emphasis on representatives of the neoavian families of Strisores. Compared with other groups of birds, members of the Strisores have very different skulls that reflect their rather specialized way of feeding. From birth, the skulls of nightbirds (e.g., nighthawks, frogmouths, etc.) and diurnal swifts have wide flat braincases, wide beaks and palates, and large orbits for locating and capturing flying insects on the wing. The shapes of adult skulls of these birds do not differ substantially from those of their youngsters. In strong contrast, the adult skulls of hummingbirds have a globular braincase and an elongated beak for probing into flowers. The skulls of baby hummers, however, more closely resemble those of swifts and the ancestors of swifts and hummingbirds than those of their adults. This indicates that, unlike other Strisores, substantial changes occur during skull development in hummingbirds. As a result, hummers now occupy a very different region of multidimensional skull shape than all other birds, including other Strisores (see figure 7). The strong developmental changes that occur in the skulls of hummingbirds have pushed them into very different niche space compared with their relatives and have allowed them to evolve a much greater range of bill shapes than most other families of birds. Again, although the genetic bases of differences in the skull morphology of nightbirds and swifts compared with hummingbirds are not yet known, these differences imply that different genetic programs might be involved in the ontogeny and evolution of these birds.

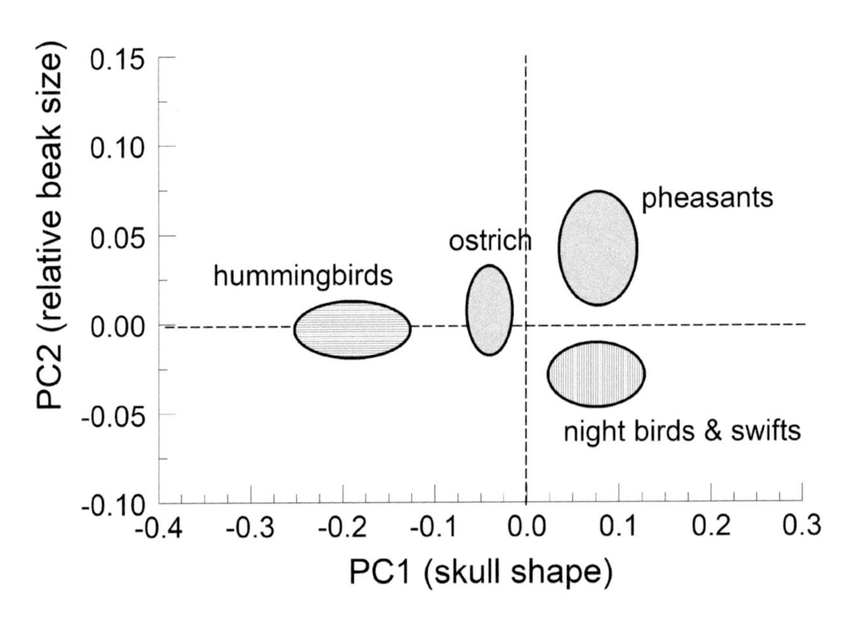

FIGURE 7 Results of an ordination (a multivariate analysis) of the skull morphology of four groups of birds, including two groups of Strisores. Based on data in Navalón et al. (2021).

Finally, we can ask, How rapidly did hummingbirds acquire their family-defining characteristics once they began to become differentiated from swifts? Which of the general scenarios found in figure 2 best represents this early evolution? Only the fossil record can truly answer this question, but hummingbird fossils, unlike those of other Strisores, are very rare. Nonetheless, the fossil hummingbird *Eurotrochilus inexpectatus* gives us a reasonable clue about this. The skull and wing skeleton of this thirty-two-million-year-old bird are basically similar to that of modern hummingbirds, which implies that the early evolution of these birds probably was rapid and likely conformed to scenario three (short fuse) of figure 2. The evolution of small size, the ability to hover, and its high size-dependent metabolic demands likely placed strong selective pressures on these birds to quickly become adept at acquiring high-energy nutrients from sources such as flower nectars. If this is true, then we might expect these birds to have evolved quickly from a proto-hummingbird morphology with a somewhat narrow bill to a more full-fledged hummer morphology. The meager fossil record suggests that this transition probably took place in just a few million years (e.g., in much less than five million years, as has also occurred in Hawaiian honeycreepers and Darwin's finches). Evolving into entirely new continental ecological niche space, as also occurs on newly formed islands, apparently favored the rapid evolution of new morphotypes in early hummingbirds. As I discuss later, this rapid evolution implies that their food plants must also have evolved rapidly to attract new kinds of pollinators.

A COMPARISON OF MODERN HUMMINGBIRDS AND SWIFTS: OR, HOW DIFFERENT HAVE THEY BECOME IN FORTY-TWO MILLION YEARS?

Comparing the basic biology of swifts and hummingbirds is like comparing apples and oranges. Although they are sister families that evolved from the same ancestral stock beginning about forty-two million years ago, today they are as different as night and day, as is apparent from information summarized in table 3. But before we examine these differences, I want to return to the morphological study published by Navalón in 2021. I do this to introduce an important statistical method called principal components analysis (PCA) that is often used to visualize morphological (or other) differences between groups of individuals (or taxonomic families, as in the present case). The aim of PCA is to reduce a dataset containing many highly intercorrelated variables into one that maximally captures the variation found in the original dataset. Variables in the reduced set are called principal components (PCs), each of which is statistically independent from the others. PCA calculations are such that PC1 captures most of the variation in the original dataset followed by PC2, PC3, and so on. In most PCA studies, it is hoped that PC1–PC3 will capture most of this variation. This is often the case in morphological studies; it is less likely to occur in ecological studies.

TABLE 3 A brief comparison of the major biological features of two sister families: swifts (Apodidae) and hummingbirds (Trochilidae)

FEATURE	SWIFTS	HUMMINGBIRDS
Taxonomic diversity	18 genera, 112 species	113 genera, 352 species
Geographic distribution	Worldwide except for Australia	Lowlands to high elevations in North, Central, and South America and the Caribbean
Size (mass) range	9–150 g	2–20 g
Morphology	Short, broad bills; wings long and pointed; feet small	Thin bills of various lengths and shapes; wings long and rounded; feet small
Sexual dimorphism and plumage	Most species are monomorphic and dull colored	About one-third of species are monomorphic and some are dull colored; all other species are sexually dimorphic with males usually having very colorful plumage
Habitats	Spend most of their lives on the wing in many different habitats	Many different habitats throughout their geographic distribution; especially diverse in the northern Andes
Diet and foraging	Feed entirely on aerial insects; constantly in flight, even at night, except to nest	Feed largely on flower nectar but also eat small insects; spend much time perching between feeding bouts
Breeding system	Monogamous with biparental care	Promiscuous breeding with males mating with several females each season

TABLE 3 *continued*

FEATURE	SWIFTS	HUMMINGBIRDS
Reproduction	Both sexes involved in all aspects of parental care; clutch size 1–5; incubation lasts 2–4 weeks; young fledged at 6–10 weeks	Only females incubate and feed young; clutch size invariably 2; incubation lasts 2–3 weeks; young fledged at 3–6 weeks
Social behavior	Long-term pair bonds in nonmigratory species; cooperative breeding in at least one species; some species are colonial roosters; often forage in mixed-species flocks	Highly territorial and socially intolerant; lek mating occurs in hermits and a few other species; some males undergo aerial courtship displays
Vocal behavior	These birds are very vocal and produce loud screams in flight	Both sexes sing insect-like songs while perched
Migratory behavior	Many species are long-distance latitudinal migrants	Many species are latitudinal migrants; some are altitudinal migrants

Sources: https://www.birdsoftheworld.org; Perrins and Middleton (1985).

In the Navalón study, ten morphological variables were measured in the skulls of thirty-six species of birds and then analyzed by PCA. Most of these species came from families of Strisores but one example each came from the evolutionarily older Palaeognathae and Galloanserae lineages for comparison. Results shown in figure 7 emphasize the large difference between the skulls of adult hummingbirds and other Strisores, including swifts. If we were to analyze the biological features summarized in table 3 by PCA, the large differences between hummingbirds and swifts would surely also be equally impressive.

As table 3 indicates, swifts have evolved an aerial lifestyle in the pursuit of insects while the lives of hummingbirds are centered on the energy produced by the flowers they visit. What could be more different for two closely related groups of birds? Despite being airborne most of the time except during nesting, some aspects of the lives of swifts are similar to those of most other birds. These include monogamous breeding systems, biparental care of offspring, and long-term pair bonds in some species. In contrast, the lives of hummingbirds differ greatly from those of most other birds. For example, they have promiscuous breeding systems in which males are only sperm donors; females are solely responsible for raising their young; and as a result, no long-term pair bonds form in these feisty nectar sippers. Another striking difference is their plumage, which is a muted dark brown in most swifts compared with the glittering beauty of many male hummingbirds. Since they are constrained to do all the incubation of their nestlings, female hummers, like those of most other birds, are usually cryptically colored as are juvenile hummers.

Another notable difference between the two families is the disparity in their diversity with hummingbirds having about three times as many species as swifts. These differences reflect in part differences in the size of their geographic ranges, which affects their chances of speciation. Many species of hummingbirds, as we've seen, are quite sedentary. They are unusual in having very small geographic ranges, especially in the Andes. As a result, their speciation potential is very high, particularly whenever gene flow among populations is low. In contrast, species of the much more mobile swifts often have large geographic ranges, which likely reduces their speciation potential, especially if gene flow among populations is high.

Other gross differences between swifts and hummingbirds are apparent in table 3, but suffice it to say that these two avian lineages have followed very different and separate evolutionary pathways once they separated in Eurasia tens of millions of years ago. Except for their long wings and small feet, these birds have very little in common today.

It's now time to examine in some detail the nuts and bolts of what makes a hummingbird. Here I intend to highlight only some of the features of hummingbirds that make them such fascinating animals. A more detailed account of many aspects of their biology can be found in the beautiful book by Glenn Bartley and Andy Swash that was published in mid-2022. The wonderful photographs of hummingbirds by Glenn Bartley and many others alone make this book a must-have for hummingbird enthusiasts.

THE BASIC BIOLOGICAL AND ECOLOGICAL CHARACTERISTICS OF HUMMINGBIRDS

AN OVERVIEW OF THE NINE HUMMINGBIRD CLADES

As we've seen, the evolution of modern hummingbirds, which began in the lowlands of South America about twenty-two million years ago (i.e., in the Early Miocene), has produced nine monophyletic clades; examples of eight of these clades are shown in figures 8 and 9. Their species richness and major features are summarized in table 4, which lists them by their ascending evolutionary age (see figure 3). Species diversity ranges from a single species (the Andean giant hummingbird) to over one hundred species in the Emeralds. Species in six of the nine clades, including the oldest ones, are medium-sized based on mass; two clades contain relatively large birds; and one clade contains relatively small birds. Large birds tend to occur in the Andes, and small birds occur in clades that initially evolved in North America (sensu lato). I'll skip the "andromorphic" topic for now and discuss it in detail below. Seven of the nine clades diversified primarily in South America whereas two of the youngest clades diversified primarily in North America (including Mexico and Central America; see table 4). Three of the clades diversified primarily in the Andes. Finally, the size and shape of bills and tails are quite diverse in these birds. Long, straight, or slightly down-curved bills characterize birds in five of the clades; bills in the remaining four clades are short to medium in length and straight. Finally, especially long or otherwise fancy (e.g., racket- or spatulate-shaped) tails occur in four of the clades. I will discuss all of these differences in detail below.

WHY IS THEIR SMALL SIZE SO IMPORTANT?

As a family, hummingbirds are the smallest birds in the world. In mass, they only range from 2 g (in Cuba's bee hummingbird) to 20 g (in the Andean giant hummingbird). I've seen many very small hummers (e.g., Calliope hummingbirds in Arizona and scintillant hummers in Panama) but the Andean giant hummer only once in the Andes of Ecuador. Whereas wingbeats of small hummers are a blur, those of the giant hummer are dramatically different: they are very slow and non-hummer-like; nonetheless, they can hover and fly backward like other hummers.

TABLE 4 Summary of the major features of the nine hummingbird clades

CLADE	NUMBER OF GENERA AND SPECIES	RELATIVE SIZE AND AVERAGE AND RANGE OF BODY MASS (G)	PERCENT OF SPECIES THAT ARE ANDROMORPHIC[A]	ELEVATIONAL RANGE AND DIVERSIFICATION CENTER[B]	OTHER FEATURES
Topazes	2 and 4	Large: 9.3 g (6.9–11.7 g)	25	Mostly lowlands; South America	Bills medium (20.8 mm; 19.9–21.7 mm) and strongly curved; Topaz males are long tailed
Hermits	6 and 36	Medium: 5.6 g (2.0–11.6 g)	15	Mostly lowlands; South America	Bills medium (31.2 mm; 21.6–40.7 mm) and straight or slightly to strongly curved
Mangoes	12 and 27	Medium: 5.3 g (3.4–7.6 g)	28	Lowlands and Andean; South America	Bills medium (21.4 mm; 13.2–35.5 mm) and slightly curved
Brilliants	13 and 51	Medium: 6.8 g (2.7–11.0 g)	17	Mostly Andean; South America	Bills medium (26.0 mm; 13.1–82.7 mm) and straight; fancy-tailed in several genera
Coquettes	18 and 61	Medium: 4.6 g (2.0–9.2 g)	20	Mostly Andean; South America	Bills small (13.7 mm; 7.9–20.2 mm) and straight; tails long and forked in several genera
Patagona	1	Large: 20.5 g	100	Mostly Andean; South American	Bill large (37.9 mm) and straight

TABLE 4 *continued*

CLADE	NUMBER OF GENERA AND SPECIES	RELATIVE SIZE AND AVERAGE AND RANGE OF BODY MASS (G)	PERCENT OF SPECIES THAT ARE ANDROMORPHIC[a]	ELEVATIONAL RANGE AND DIVERSIFICATION CENTER[b]	OTHER FEATURES
Mountain Gems	7 and 14	Medium: 6.5 g (5.4–9.1 g)	13	Mostly lowlands; North America	Bills medium (25.0 mm; 19.2–34.3 mm) and straight
Bees	15 and 36	Small: 3.0 g (2.2–4.1 g)	3	Lowlands and mountains; North America	Bills small (15.9 mm; 12.0–20.2 mm) and straight
Emeralds	27 and 108	Medium: 4.6 g (2.5–10.7 g)	14	Mostly lowlands; South America	Bills small to medium (19.3 mm; 12.0–28.2 mm); tail long and forked in *Eupetomena*

Sources: Diamant et al. (2021), McGuire et al. (2014), and Rodríguez-Flores et al. (2019).
[a]Andromorphic indicates adult females that have male-like plumage.
[b]North America as defined here extends to Panama; South America extends south of Panama.
Note: Morphological data are courtesy of F. G. Stiles.

Why is body size important? As many biologists know, size can be destiny because many (most?) of the biological and ecological features of Earth's organisms are size dependent. In birds, for example, the following features are *positively* correlated with body mass: incubation period, heart mass, brain mass, skeletal mass, territory or home range size, and life span. Features that are *negatively* correlated with body mass include resting heartbeat, wingbeat, resting breathing rate, and population density. One important life history feature that is not size dependent in birds is clutch size. Thus, hummingbirds, many doves, and vultures all typically produce two-egg clutches.

Of critical importance for hummingbirds is the negative relationship between mass-specific metabolic rate and body mass in birds and mammals. Being the smallest birds, hummers are "stuck" with the highest mass-specific metabolic rates of any vertebrate. As a result, their energy needs can be extremely high, and they need to eat a lot each day.

FIGURE 8 Four hummingbird clades I: A. Topaz (white-necked Jacobin, *Florisuga mellivora*), B. Hermit (long-billed hermit, *Phaethornis superciliosus*), C. Mango (green violet-ear, *Colibri thalassinus*), and D. Coquette (Tyrian metaltail, *Metallura tyrianthina*). Photo credits: Ted Fleming.

THE CONSEQUENCES OF BEING AN ENDOTHERM

As we all know, birds and mammals differ from other vertebrates (and life in general) in being endothermic (or in the vernacular, warm blooded). This means that their body temperatures are determined by their internal metabolism rather than by external abiotic factors such as air or substrate temperatures as in ectotherms such as reptiles and the rest of the living world. During the day, many hummingbirds are operating at a body temperature of about 41°C, a value that is a bit higher than most other birds and mammals. It takes lots of energy, probably over 90 percent of the energy they acquire from their

FIGURE 9 Four hummingbird clades II: A. Brilliant (Talamanca hummingbird, *Eugenes spectabilis*), B. Mountain Gem (purple-throated mountain gem, *Lampornis calolaemus*), C. Bee (Calliope hummingbird, *Stellula calliope*), and D. Emerald (broad-billed hummingbird, *Cynanthus latirostris*). Photo credits: Ted Fleming.

food each day, to maintain this high body temperature. That's the price endothermic birds and mammals pay for having high body temperatures that are independent of outside temperatures. One of the benefits of this, however, is the ability to live in a wider range of habitats than many ectotherms. For example, hummingbirds have become important pollinators of plants growing at high elevations (up to about 5,000 m) in mountains where daytime temperatures are sometimes too low for bees and many other insects to be active. The evolution of a high diversity of high-elevation Andean hummingbirds reflects this.

HOW HIGH ARE THE METABOLIC RATES OF HUMMINGBIRDS COMPARED WITH OTHER BIRDS AND MAMMALS?

Simply put, their mass-specific metabolic rates (as measured by rates of oxygen consumption) during hovering and forward flight are the highest known for birds and mammals. Raul Suarez, a physiologist and expert in vertebrate metabolism, has proposed that these high rates set the lower limits on body size and upper limits on aerobic metabolic rates of any endothermic animal. While at rest, the mass-specific rate of oxygen consumption in (small) hummingbirds is about 4 mL O_2 per gram per hour; it increases nearly tenfold during hovering flight. Additional metabolically related features of these birds are also superlative. At rest their heartrates are about 500 per minute and increase to about 1,300 per minute during hovering; to accomplish this, their hearts are about twice as large as other birds of similar size. While feeding, their tongues move in and out about thirteen times a second. And to accomplish their hovering flight, their flight muscles represent about 30 percent of their body mass compared to much less than 25 percent in other birds.

A long-standing question has been, how do hummingbirds fuel their high metabolic rates? Potential fuels include recently ingested nectar sugar or stored lipids (fats) that have been produced by the conversion of sugars in their livers. Recent studies using carbon stable isotopes have shown that hummingbirds directly oxidize nectar sugars to fuel their diurnal metabolism. As a result, the rate of sugar flux through their bodies is extremely high—about fifty-five times higher than that of nonflying mammals. And to digest and direct carbohydrate-based fuels to their flight muscles, they have evolved a highly efficient form of the enzyme sucrase located in the small intestine that helps them to quickly convert nectar sugar into high-energy molecules that fuel their muscles during flight.

In contrast, at night and during their energy-intensive migrations, hummingbirds use stored lipids for fuel. During migration, these birds feed intensely during stopover rests and build up their fat stores by 25–40 percent in just four days. The ability of hummingbirds to quickly produce and then use lipids for their metabolism suggests that, like their enzyme sucrase, the enzymes involved in lipogenesis (fat production) likely differ from those of other birds. A recent detailed genetic analysis of the enzymes involved in lipogenesis in hummers compared with other bird (including the chimney swift) and human genomes indicates that these enzymes are indeed very different from those of other birds. This difference likely reflects an adaptation for rapid lipogenesis that is unique in fasting and migrating hummingbirds—another example of how exceptional hummingbirds are compared with other birds.

Taken together, the overall suite of adaptations involved in a hummingbird's high-energy lifestyle is extremely impressive. The following quote, taken from Suarez (1992, 568), highlights these adaptations, all of which differ significantly from other birds and mammals of similar size: "Larger and faster hearts, high hematocrits [red blood cell counts] and O_2 unloading efficiencies, more capillary surface area per muscle fiber surface areas, greater mitochondrial volume densities, . . . more enzyme copies in pathways of glucose and fatty acid oxidation and synthesis, and more absorptive intestines, have all formed part of a concerted, integrated pattern" that has greatly increased a hummingbird's metabolic capacity.

It is important to remember that the features that make hummingbirds metabolically and morphologically unique among birds did not evolve de novo. Instead, they evolved via strong directional selection from ancestors equipped with similar metabolic and morphological attributes but whose performance was less spectacular than that of hummingbirds. That is, natural selection in hummingbirds has favored genetic variations in particular chemical and morphological traits (e.g., enzymes, red blood counts, muscles, intestines, livers, hearts, and lungs) that improved the ability of these birds to extract and use energy from flowers more efficiently. Ancestrally, swifts and several night birds (Strisores) have metabolic rates that are lower, rather than much higher as in hummingbirds, than expected for birds of their size, and their normal body temperatures are several degrees lower than those of hummingbirds. It thus appears that nonhummingbird members of Strisores are metabolically much more conservative than hummingbirds, which supports the idea that selection has favored higher rates of metabolism during the evolution of these nectar feeders.

To reduce the overall cost of their high-energy lifestyle, hummingbirds often undergo torpor at night. While in torpor, their body temperatures decrease by at least 8°C, and their heart rates and rates of respiration and metabolism also decrease substantially. Their metabolic rates decrease to about 10 percent of their normal resting metabolic rates—a considerable energy savings. Their small sizes are advantageous when hummingbirds arouse from torpor. It takes them only an hour or so to do this compared with several hours for larger birds. The use of torpor to save energy is generally not common in birds but does occur occasionally in a few swifts, nightbirds (Strisores), and swallows, particularly when abundance of their prey declines. So, it is likely that the physiology of their ancestors preadapted hummingbirds to use torpor as an energy-saving adaptation.

Although the use of daily torpor is generally common in hummingbirds, it does vary among species based on their body size and phylogeny. An analysis of the use of torpor in thirty-one species from all of the major clades of hummingbirds revealed that nightly torpor is nearly obligatory in small species (i.e., < 6 g) whereas it is more facultative in larger species (i.e., ≥ 6 g) where it tends to be employed only under particularly stressful conditions of temperature and food availability. Furthermore, the use of torpor in high-Andean hummingbirds has a significant phylogenetic component. Coquettes, which are well represented among Andean species, undergo deeper and longer bouts of torpor than other Andean hummers.

Finally, to put all of this metabolic information into the context of whole animal biology, we can ask, What is the daily cost of living, as measured in kilojoules (kJ) of energy (1 kJ = 0.24 kilocalories), of a typical Temperate Zone hummingbird (e.g., Anna's hummingbird weighing 4.5 g and living in southern California)? How much energy does this species need to acquire from its environment every day to stave off Schrödinger's pull of entropy (see page 4)? This cost is currently expressed as field metabolic rate (FMR), and it can be measured by a process called the doubly labeled water technique. This technique involves injecting a small amount of an aqueous solution containing stable (i.e., nonradioactive) isotopes of hydrogen and oxygen into a free-ranging animal; a baseline blood sample is then taken before the animal is released to resume its daily activities. After it is recaptured, another small blood sample is removed and analyzed to determine how much CO_2 (a product of metabolic activity) and H_2O have been released between captures. An animal's FMR can be determined from the CO_2 data; its rate of water turnover can be determined from the H_2O data. Measured in this way, the FMR of a typical-sized hummingbird—the 4.5 g Anna's hummingbird—is 32 kJ per day. And that of the 20 g Andean giant hummingbird is a whopping 211 kJ. I'll translate these values into the number of flower visits these birds need to make to acquire this amount of energy a bit later. But the short answer is lots.

It isn't obvious from these estimates of the FMR of two species of hummingbirds how expensive their daily costs of living really are. Some comparative data are needed for this. So let's compare these values with estimates of the FMRs of a 4.5 g and a 20 g rodent—a standard comparison in the physiological literature. We can do this based on the following mathematical relationship: FMR in kJ = aX^b where a equals an empirically determined constant (the value of Y when X is zero), X equals an animal's mass in grams, and b equals the slope of the regression line relating FMR (Y) to mass (X). This equation is known as an *allometric equation* in which the slope b describes how FMR (or some other variable Y) changes as a function of an animal's mass (X). Allometry thus deals with the biology of scaling (e.g., how does FMR scale with mass in a particular species or group of animals?). It is a common statistical tool in comparative physiology studies (among many others). This slope can take one of three possible values: if $b = 1.0$, then the relationship between Y and X is called *isometric*; a unit change in X results in a unit change in Y, producing a straight-line relationship between X and Y. If b is < 1.0, the relationship between Y and X is called *negatively allometric*; Y changes more slowly than X and the relationship is curvilinear decreasing (i.e., concave down). And if b is > 1.0, this relationship is called *positively allometric*; Y changes faster than X and the relationship is curvilinear increasing (i.e., concave up).

Based on a lot of empirical data, we know that for hummingbirds, the allometric equation relating FMR to mass is FMR = $5.61X^{1.21}$. Since $b > 1.0$, this relationship is a *positive* allometric one. FMR increases faster than mass. For comparison, the allometric equation for rodents is FMR = $5.48X^{0.712}$, and the relationship is a *negative* allometric one. That is, FMR increases more slowly than mass as mass increases. Plugging the mass values of 4.5 g and 20 g for two rodents into the rodent equation, their FMRs turn out to be 16.0 kJ and 46.3 kJ, respectively—one-half to one-fifth of the hummingbird values. For an even more dramatic comparison, consider the FMR of a 4.5 g or 20 g iguanid lizard based on their allometric equation of FMR = $0.301X^{0.793}$—another negative allometric relationship. These FMRs turn out to be 0.99 kJ and 3.24 kJ, respectively. The daily cost of living for the lizard, which is an ectotherm, is only a tiny fraction of that of a similar-sized mouse or hummingbird, both of which are endotherms.

When one scans the FMR data assembled for many species of mammals, birds, and reptiles by Ken Nagy and colleagues in 1999, it turns out that virtually all of the allometric relationships, with one glaring exception, are negative (i.e., $b < 1.0$). That exception, of course, is hummingbirds, in which FMR is positive allometric. And why is the cost of living so high in hummingbirds? It's because they hover to obtain nectar from flowers. Hovering turns out to be an extremely expensive form of locomotion. In the Andean giant hummingbird, for example, its FMR (i.e., its daily cost of living) is about 4.9 times greater than its minimum metabolic rate when it's resting (i.e., its basal metabolic rate or BMR). But when it's hovering, its metabolic rate is about 9.1 times its BMR. Nonetheless, despite the high energetic cost of hovering, this is the adaptive route that these birds have taken to be able to harvest minute amounts of energy from most of the flowers they visit. As we will see, certain nectar-feeding bats also hover when visiting flowers, but the cost of hovering in these mammals is not nearly as high as it is in hummingbirds, and FMR values for bats decrease with body mass, as is true in most other endotherms.

An obvious way for a hummingbird to reduce its daily energy expenditure is to minimize the time it spends hovering at flowers. When I watch hummers feeding at flowers or at the hummingbird feeders in our yard, I note that they spend much more time sitting on a perch rather than hovering. Others that have studied their feeding behavior in the field report that hummers spend about 20 percent of their day foraging of which about 75 percent is spent simply resting. Most of that time is probably spent shunting the sugar they've just ingested from their crops (an expanded region of their esophagus) into their intestines and then, via their circulatory system, to their flight muscles where it is converted into high-energy molecules. This process obviously saves energy and reduces the time they have to spend hovering at flowers.

HOW HUMMINGBIRDS MEET THEIR OXYGEN NEEDS

Hummingbirds cannot live on energy alone. Their high activity levels and metabolic rates demand that they constantly acquire lots of oxygen to fuel their aerobic metabolic machinery. During flight, their rate of oxygen consumption is about twelve times higher than it is during rest. It is ten times higher than a human during strenuous exercise. How do they do this? Do they possess unique anatomical and physiological methods for meeting this need? Studies of oxygen consumption and its delivery throughout the body in hummingbirds have indicated that they are indeed exceptional in doing this. For example, a four-gram hummingbird has about *twenty-five kilometers* of capillaries in its flight muscles compared with about ten kilometers in a similar-sized mammal. Delivery of large amounts of oxygen to their flight muscles involves a number of crucial adaptations, including a high capacity of their lungs to allow oxygen to diffuse rapidly into the bloodstream, high cardiac output from their large hearts to distribute oxygen throughout the body, a high ratio of capillaries to muscle fibers in flight muscles, a high density of mitochondria (the power-generating factories) in flight muscles, and a high concentration of metabolic enzymes in flight muscles. All of these features and processes greatly exceed expectations for birds or mammals of their size. And all of these actions require a coordinated series of events controlled by a suite of regulatory genes that turn on and off at just the right time. So next time you see a hummingbird hovering at a flower or feeder, take a minute to marvel at all of the biological activities that are going on inside that little body. It's truly amazing, right?

THE CHALLENGES OF LIVING AT HIGH ELEVATIONS

The elevational range of hummingbirds is impressive—from sea level to over five thousand meters. As we've seen, a substantial number of species, especially in the Coquette and Brilliant clades, have radiated extensively in the Andes. What kinds of adaptations do we find in the high-elevation species that are lacking in lowland species? Before moving into the mountains, let's first review the basic features of Hermits, a clade that has not been able to successfully colonize high-elevation Andean habitats, and see how they differ morphologically from other hummer clades. Although a few species of Hermits reach elevations of up to two thousand meters, these birds are basically residents of lowland wet tropical forests in South America. Gary Stiles conducted an extensive morphological analysis of females (to avoid sexually selected traits of males) of 21 species of hermits and 115 species of non-hermits (the other eight clades) and summarized these data via discriminant analysis, a multivariate statistical method similar to principal components analysis (PCA) that I described on pages 34–35. As seen in figure 10, the morphological space occupied by the two groups is nearly non-overlapping. Hermits differ morphologically from non-hermits most strongly in having longer and thicker bills, smaller feet, and longer tails. The two groups do not differ in body mass and various wing parameters. He then looked for correlations between each species' morphological measurements and its average elevation. He found significant correlations (either positive or negative) for eight of nine variables in non-hermits but no significant correlations in hermits. For example, body mass, wing length and width, and foot size increase and bill length decreases with elevation in non-hermits whereas none of these trends occurs in hermits. In other words, elevation matters in the morphology of non-hermits but not in hermits. As Stiles writes, "It is as though the hermits have an essentially invariant morphological 'package' that is resistant to change; they move upslope to the limits established by this 'package,' and no further." In this, they resemble how Hawaiian thrushes are morphologically invariant compared with Hawaiian cardueline finches that I discussed on page 35. One reason why hermits don't occur at high elevations is that their major food plants, the *Heliconia* megaherbs, also do not occur at high elevations because they are intolerant of cold temperatures. A close evolutionary association between these birds and plants has apparently helped to restrict their distributions.

Let's now review another study by Gary Stiles of the morphological features of hummingbirds that have evolved extensively at high elevations in the Andes. These environments can be challenging for small, high-energy birds because of their low temperatures, low air densities, and reduced oxygen availability. How do the morphologies of high-elevation species compare with their low-elevation relatives? Using the database in his previous morphological study, he found that two clades—Coquettes and Brilliants—contain the

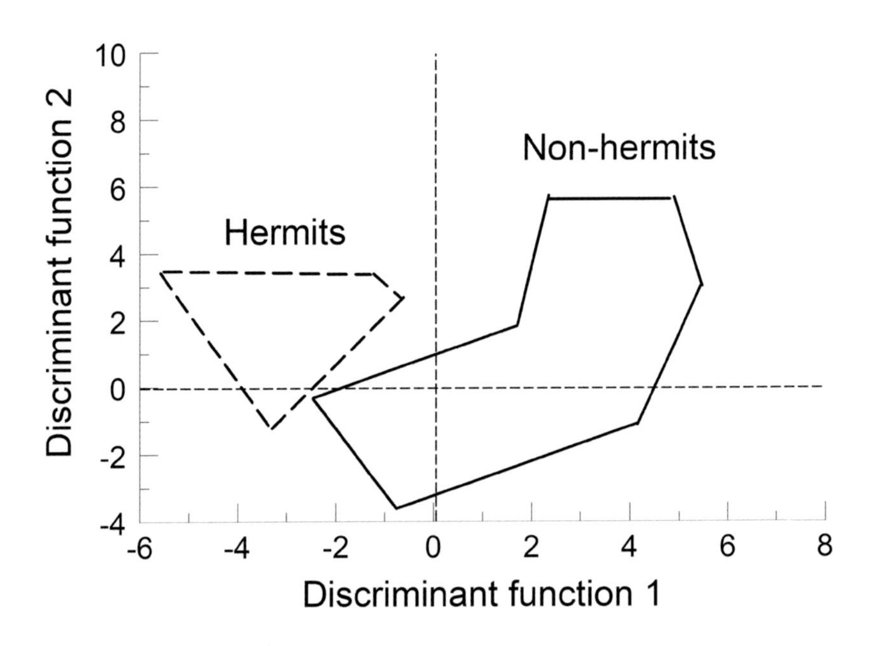

FIGURE 10 A 2D multivariate ordination of the morphology of hermits vs. non-hermit hummingbirds. Based on data in Stiles (2004).

greatest number of species that occur at elevations of three thousand meters or more. But species in these two clades have not converged to produce a single "high-elevation morphology." Instead, they differ in many morphological features. Thus, both high- and low-elevation Coquettes are relatively small and have short, straight bills, broad wings, and large feet. In contrast, both high- and low-elevation Brilliants are relatively large and have long straight bills, relatively narrow wings, and small feet. Within these two clades, therefore, no conspicuous morphological differences exist between low and high-elevation species. High-elevation Coquettes occur more often in open paramo habitats whereas Brilliants occur more often in closed forests. Similarly, their different bill sizes reflect differences in the kinds of flowers members of these clades visit at high elevations. Coquettes are strongly associated with the flowers of composites (Asteraceae) whereas Brilliants are strongly associated with tubular flowers of the blueberry (Ericaceae) and other families. Finally, the large feet of Coquettes allow them to perch, rather than hover, at flowers, whereas most Brilliants hover at flowers. Members of other clades that occur at high elevations have tended to converge morphologically with these two ecomorphs: either the Coquette morph or the Brilliant morph. This trend supports the idea that the adaptive radiation of high-elevation hummingbirds has occurred in two distinct ecological niches: (1) open habitats containing small flowers that occur in dense inflorescences produced by shrubs and (2) closed forests containing many shrubs and epiphytes that produce large tubular flowers.

Since oxygen levels are significantly lower at high elevations, as I've experienced by huffing and puffing at elevations of four thousand to five thousand meters in the Andes, it is of interest to know what kinds of morphological and physiological adaptations these high-energy birds possess at high elevations. To my knowledge, however, studies comparing the physiology of low- and high-elevation Coquettes and Brilliants have not yet been done. Since we already know that the morphologies of these birds differ along elevational gradients, we should also expect to find significant differences in their behavior and physiology along these gradients. For example, energy conservation should be very important, perhaps expressed by less hovering and a greater use of torpor, at high elevations.

A hint at the constraints imposed by high elevations and their reduced oxygen levels and air density comes from a laboratory study of the response of low-elevation ruby-throated hummingbirds (a Bee) to air mixtures in which the partial pressure of oxygen and air density were manipulated by the addition of helium to the air in test chambers. Reduced oxygen challenges the ability of these birds to undergo intense exercise such as hovering; reduced air density challenges their ability to maintain power output during flight. Results indicated that in mixtures in which both oxygen content and air density were reduced, birds were unable to hover when oxygen levels dropped to about 12 percent (the equivalent of an elevation of about 4,000 m) and air density was 63 percent of normal. Thus, both the metabolic physiology and wing size and shape of these low elevation hummers would have to change for them to live at high elevations.

HUMMINGBIRD WINGS AND HOW THEY HOVER

Although a number of different birds can hover in place for brief periods of time; hummingbirds are clearly the champion hoverers among birds. As I've mentioned, they typically spend about 20 percent of their day foraging, which can involve hovering at flowers several times a minute until their crops are full, and it has been reported that captive hummingbirds can hover nonstop for up to one hour at a time. What is the morphological machinery that allows them to do this? Modifications of their skeletons associated with their unique mode of flight include a greatly enlarged sternum for the attachment of their large flight muscles; eight pairs of ribs rather than six as in other land birds, for protecting the ribcage from the stresses of vigorous flight; a strong coracoid bone with a unique ball-and-socket joint that connects the wing to the sternum; and a highly modified wing featuring a short, dumbbell-shaped humerus (as seen in *Eurotrochilus*), a short and fused radius and ulna, and a greatly elongated series of fused hand bones that support the primary flight feathers (figure 5). The fused skeletal segments of the radius, ulna, and hand bones can rotate as a unit during flight, changing the angle of attack of their wing feathers.

This series of skeletal and muscular features allows hummers to hover as well as fly forward and backward and to produce lift during both the wing's upstrokes and downstrokes—feats that are unique among birds. High-speed cinematography, pioneered by the polymath Crawford Greenewalt (1902–93), a chemical engineer and president of Dupont Company from 1948 to 1962, and more recent X-ray videography have revealed the mechanics of this unusual ability. These studies indicate that hummingbird wings operate like a variable-pitch rotor in which the wrist and forearm are inflexible and act as a single

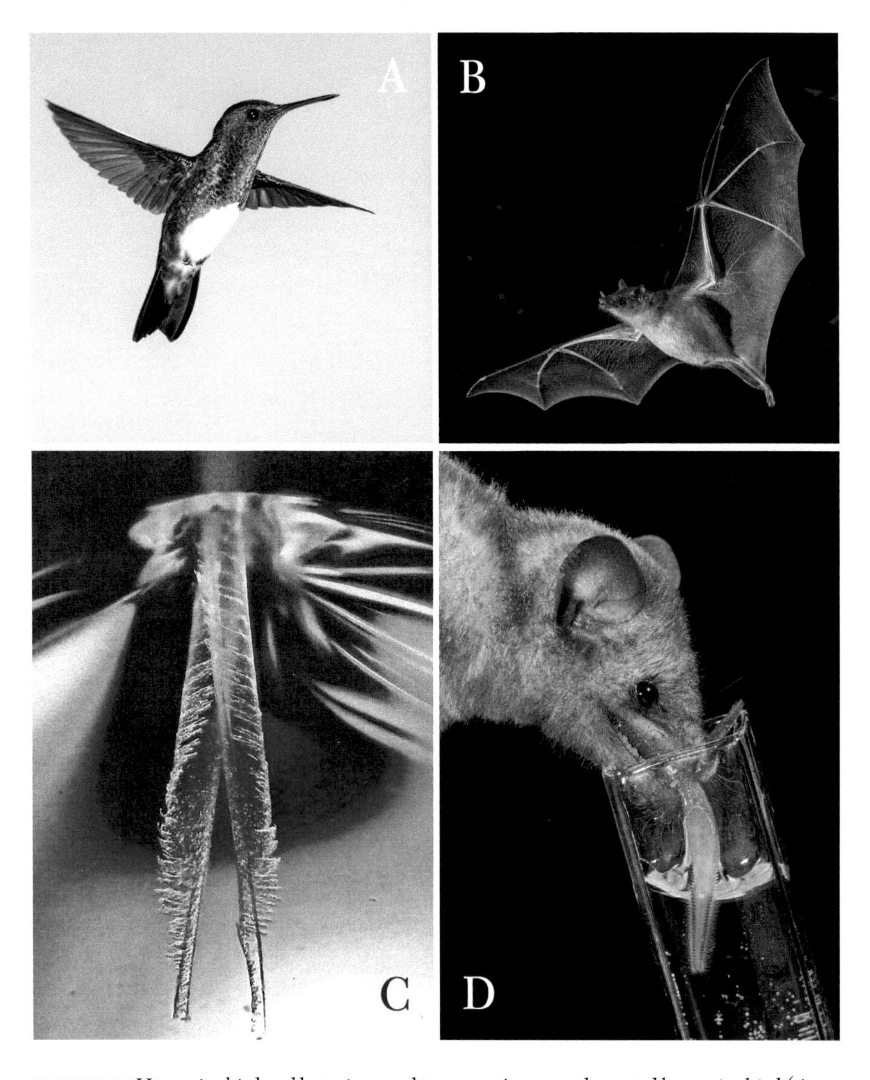

FIGURE 11 Hummingbird and bat wings and tongues: A. snowy-breasted hummingbird (*Amazilia edward*), B. lesser long-nosed bat (*Leptonycteris yerbabuenae*), C. hummingbird tongue, and D. nectar bat tongue. Photo credits: Ted Fleming (A, B); C courtesy of A. Rico-Guevara; and D courtesy of M. B. Fenton.

unit like an oar (figure 11). As in other birds, during forward flight the path of a hummer's wing tip traces a vertical oval whereas, unlike other birds, during hovering it moves horizontally in a figure eight. By tilting its wings slightly up while hovering, these birds can easily move up and back; by tilting them slightly down, they can move slowly forward. Powering this flight style are two especially large sets of muscles: the pectoralis (chest) muscles for the downstroke and the supracoracoideus (shoulder) muscles for the upstroke. Both of these muscles in hummingbirds are proportionately much larger than in other birds.

Finally, in addition to feeding via hovering at flowers, hummingbirds display an amazing variety of flight maneuvers whenever they pursue insects, evade predators, and interact with each other during territorial chases and courtship displays. Hummingbirds are thus not just built for hovering; they are more generally built for very fast, maneuverable flight, including tight turns, pitches and rolls, and fast ascents and descents.

Hummingbirds vary in size (mass) and the size and shapes of their wings, and these features affect their flight performance. Their mass varies by a factor of ten, and some species have longer and narrower wings than others. Species with long narrow wings are very agile fliers whereas those with wider wings are slower but more maneuverable fliers. Statistical analysis of many species of hummingbirds indicates that differences among species in mass and wing size and shape are highly correlated, which indicates that these traits vary as adaptive units, as is the case in their highly integrated skull characteristics discussed above (pages 30–31). Furthermore, when phylogenetic relationships are taken into consideration, closely related species are more like each other in their morphology and flight behavior than expected by chance. Thus, lowland species tend to be smaller in mass and wing area than high-elevation species and are more agile in flight. Upland species need larger wings to deal with the lower air density at high elevations. Overall, however, despite differences in size and their wings, the flight characteristics of all hummingbirds are actually quite similar compared with, for example, other kinds of insectivorous birds and bats. This suggests that the evolution of hummingbird flight machinery has occurred within very narrow bounds dictated by their small size, their high metabolic demands, and the mechanics of extracting energy efficiently from very stingy flowers under a variety of environmental conditions that range from sea level to over 5,000 m above sea level.

HUMMINGBIRD FEATHERS: ALL THAT GLITTERS . . .

Feathers are probably the most important defining characteristic of birds, although they were present in their modern form in some dinosaurs before the evolution of true birds. As in many other features, hummingbirds differ from other birds in some of their feather characteristics. For example, their tiny bodies are covered by about 1,200 feathers at a density that is much higher than in other birds. In addition, although the plumage of some juvenile hummers includes down feathers for insulation, down feathers are missing in adult plumages, perhaps as an adaptation for maximizing their ratc of heat loss generated during their foraging and other high-energy activities.

We also tend to think of all hummingbirds as glittering green jewels with males being gaudier than females, but in actuality their plumage is quite varied among species (figures 8 and 9). About one-third of hummingbird species, for instance, are sexually monomorphic (i.e., the plumage of males and females is similar). These species include most of the brown-hued Hermits, which live in the dark understories of tropical and subtropical forests, and the Andean giant hummingbird. Not all of the monomorphic kinds, however, are dull colored. Species of Emeralds and violet-ears (Mangoes), for example, are brilliant green in color. But the majority of species are sexually dimorphic with females being far less gaudy than males. Males in these species often have patches of brilliant green, blue, or purple (and sometimes red or yellow) feathers on their heads, throats, and upper chests; some are also adorned with ear tufts, crests, or very long tails. In addition, in about one-quarter of hummingbird species, some females have male-like plumage, suggesting that the evolution of plumage characteristics in hummers can be complex. Thus, whereas the underlying flight machinery of hummers is very similar among species, their external trappings are anything but similar both within and between species, as are their bills which I'll discuss later.

Hummingbirds, of course, are not the only colorful birds. One only has to think of parrots and toucans among nonpasserines and tanagers and wood warblers among passerines to realize this. But the brilliant iridescent feather colors of hummingbirds arise in a way that differs from most other birds. Plumage colors in most birds are the products of organic compounds deposited in feather cells that absorb and reflect different wavelengths of the visible electromagnetic spectrum, which ranges from 300 (blues) to 750 (reds) nanometers. The major organic compounds include melanins (earth tones and black), carotenoids (bright colors such as yellows, reds, and purples), psittacofulvins (the bright colors found only in parrots), and porphyrins (the bright colors of turacos of Africa and a few others). Carotenoids of over three dozen different kinds are interesting because they are derived from plant products and are acquired and modified chemically by birds from their diets.

In contrast, the highly iridescent colors of parts of the plumage of hummingbirds and a variety of other birds are called *structural colors* that result from optical interactions between incident light and minute structural elements called nanostructures within their feather barbules (small divisions of each feather that link a feather's barbs together to create a smooth but flexible surface). The degree of iridescence changes in color with the angle of observation and degree of illumination. So hummingbird iridescence turns on and off depending on light conditions. Under full sunlight, for instance, the head and small gorget of males of Anna's hummingbird are brilliant red whereas they are dark colored in the shade. A recent sophisticated analysis of the range of colors found in hummingbird plumages indicates that this array exceeds that of all other birds. Colors found in the UV and blue portion of the electromagnetic spectrum—colors that we cannot see but hummingbirds can—are especially common in hummingbirds. As I discuss in more detail below, the visual system of hummingbirds is extremely impressive. But perhaps more impressive is the fact that it has evolved from nocturnal ancestors, which have very limited color vision owing to a loss of color-detecting opsin genes in their retinas.

In addition to describing in detail the basic flight characteristics of hummingbirds, Crawford Greenewalt also worked out the physics of this iridescence. He noted that each thin barbule contains several to many layers of tiny melanin-containing nanostructures (also called melanosomes) that are sometimes filled with air. When incoming light is reflected back after passing through these structures, it is bent (refracted) in a process known as coherent scattering (rather than chaotic scattering) that produces different pure colors depending on their refractive index and the angle at which the scattered light is being observed. The refractive index of pure white is 1.0, and the refractive indices of other colors of the electromagnetic spectrum range from 1.5 (blue) to 1.85 (red). A familiar example of this is the refraction of light passing through rain drops that produces the colors of rainbows. Similarly, the variety of hues that we see in the plumage of hummingbirds consists of colors that are being reflected from their feathers after light has passed through and been refracted by the nanostructures. The number, size, and arrangement of nanostructures within barbules ultimately determines the nature of this scattered light. Plumage iridescence in these birds is generally limited to their body feathers and does not occur on their wing and tail feathers.

We often conclude that males that are brightly colored, heavily armored, or larger than females are the products of sexual selection in which males compete among themselves for access to sexually receptive females. In addition to superior competitive ability in males, sexual selection also involves nonrandom mate choice in females, which can reinforce differences in competitive abilities among males. So the question becomes, are the brilliant plumages of many male hummingbirds the product of sexual selection? The standard answer to this question is, yes, sexual selection is involved in plumage evolution in hummingbirds. But there appears to be more than just sexual selection going on here.

For example, on the female choice side, this process likely stems from the fact that, unlike their sister family the swifts, color is a crucial aspect of the visual world of hummingbirds. After all, finding appropriate flowers to visit certainly involves paying attention to colors. Thus, females are likely to have a sensory bias toward color in their choice of mates because of the overall importance of color in their lives. Females seem to prefer bright-colored males possibly because they contain "good genes," or are healthier, or are superior competitors for resources. But tests of these possibilities have not yet been done. And the absence of bright plumage in hermit hummingbirds, whose males are well known for being highly competitive for mates (as discussed below), suggests that sexual selection may not necessarily be involved across the board in the evolution of male plumages in hummingbirds.

One possible explanation for this evolution that doesn't rely solely on sexual selection is Richard Prum's discussion of the importance of female mate choice in *The Evolution of Beauty*. In it, he suggests that the bright colors of many birds, among a whole host of other anatomical features of animals, simply exist because they are beautiful. They don't serve any other function that enhances fitness or adaptation other than being beautiful. This, of course, begs the question, how is beauty defined operationally in animals? By what mechanism is beauty identified in animals and what is its genetic basis? Clearly, given the high diversity of plumage colors in birds, not to mention the "beautiful" features of many other animals (e.g., coral reef fish and butterflies), it seems highly unlikely that a universal definition of "beauty" exists in the animal kingdom. So, how can it be defined operationally and selected for in any given group such as hummingbirds? At the very least, it must be defined based on the visual system of the species of interest and not on our visual and cultural systems. So the question remains, do hummingbirds consider themselves to be beautiful?

Another aspect of hummingbird plumage evolution that suggests that it can't evolve solely via sexual selection is the phenomenon of *androchromatism* or female-limited polymorphism. That is, a significant proportion of females in some species have male-like plumage whereas others have female plumage. Thus, instead of just two plumage categories (i.e., both sexes monomorphic or dimorphic) occurring in hummingbirds, three categories exist (monomorphic, dimorphic, and andromorphic). Diamant and colleagues have reviewed this topic in hummingbirds and note that about 44 percent of dimorphic species appear to contain andromorphic females. Although these species can be found in most hummingbird lineages (clades), they are especially common in Topazes (the ancestral clade) and Mangoes (another early clade; figure 3) in which 25–28 percent of species contain andromorphic females (table 4). Their analysis further suggests that this condition has evolved independently at least 28 times in hummingbirds, indicating that andromorphism is an evolutionarily labile trait.

One reason why male-like plumage might be selected for in females has to do with social, rather than sexual, factors. Hummingbirds are highly territorial animals, and males are well known for aggressively defending food resources against members of their own and other hummingbird species. Within species, males are particularly aggressive toward females in dimorphic species. But observations of social interactions at food sources in the andromorphic white-necked Jacobin (a Topaz, figure 8A) in Panama have shown that male-like females are attacked less frequently than other females. So being andromorphic possibly gives females a feeding advantage over other females, which begs the question, why be dimorphic in the first place? Why aren't there more monomorphic species of hummingbirds? The answer is likely to involve the advantages to females of being less conspicuous than males during the nesting season when they are incubating eggs and brooding young. For many female hummingbirds, avoidance of being detected by visual predators while nesting, as is the case in many species of birds, would seem to be as important (or perhaps more important?) as not being hassled by males at food sources.

Finally, in addition to advertising themselves with brilliant plumage and vocalizations, some male hummingbirds use mechanical sounds produced by their tail or wing feathers to make their presence known, especially during courtship. This is particularly common in species in the advanced Bee clade in which many males produce conspicuous sounds during their courtship dives. I hear males of this clade—Anna's and Costa's hummers—making these sounds above my yard in the spring. The bell-like tinkling sounds made by the wings of males of another member of this clade—broad-tailed hummers—are common throughout the mountains of western North America during the summer. Sophisticated analyses of hummingbird tail feathers using a scanning laser Doppler vibrometer (pretty high tech!) in a wind tunnel have revealed that air passing over them causes aeroelastic flutter which in turn produces audible sound. The faster these feathers vibrate, the louder the sound. Adjacent tail feathers can either vibrate at the same or different frequency. When the latter happens, complex sounds are produced. Not surprisingly, shape of the tail feathers, which varies among species, affects the nature of the sound it produces. Researchers conducting this study concluded that the tails of Bee hummingbirds function as acoustic organs and that they have evolved via sexual selection resulting from female preferences for certain sounds during courtship displays.

Not answered in this study is why acoustic signaling apparently occurs in only one of the nine hummingbird clades. A hypothesis to explain this might be that Bee hummingbirds are common in open habitats where elaborate courtship displays involving rapid ascents and dives by males are possible. These elaborate displays are not possible in the cluttered understory of lowland tropical forests, the principal habitat of hermit hummingbirds. Once males began to display in this manner in open habitats, then any sounds produced by their tail feathers would further capture the attention of females, and selection for feathers that produce particular sounds would ensue. A geographic analysis of the major habitats (closed or open forest) of the different hummingbird lineages confirms that Bees are primarily open habitat species whereas Hermits, Brilliants, and Mountain Gems are primarily closed forest species.

A TALE OF HUMMINGBIRD TAILS

Acoustic considerations aside, the topic of hummingbird tails—their diversity and evolution—is another fascinating aspect of hummingbird biology that has attracted considerable scientific attention. Those of us living in North America only see the tip of the hummingbird tail iceberg because the species around us all have short, fan-shaped or rounded tails (compare figures 8 and 9). Although the number of feathers in hummingbird tails is very consistent throughout the family, the shape of some of those feathers, especially in the males of tropical species, is very diverse: they can be forked, scissorlike, wedge-shaped, spatulate, graduated, or pointed. All of this variation raises two basic questions: is this morphological diversity another product of sexual selection and how does tail size and shape affect the cost of flight?

From an aerodynamic point of view, the tails of birds serve three basic functions. First, they help maintain aerodynamic stability like the wings and tails of airplanes. Second, they can help control the angle of attack of the wings to reduce the pitching movements they create. I see this all the time when hummingbirds visit my feeders. Their tails are often bobbing back and forth as they hover in place. Third, by acting like a rudder, they can provide important lift during acceleration, turning, and slow flight. In addition, tails and the air flowing over them, like wings, produce lift. Long tails can produce more lift than short tails and they assist in rapid turns, but they do this at a greater energetic cost to their tail muscles. Another cost is the increased drag associated with long tails. Putting all of this together, studies dealing with the aerodynamics of bird flight suggest that the optimal tail design (i.e., one the maximizes lift and minimizes drag) involves tails in which the outer tail feathers are slightly longer than twice the length of the inner tail feathers as seen in many kinds of fork-tailed birds (e.g., scissor-tailed kites, some swifts, and many swallows). But optimal tail design clearly depends on a bird's lifestyle. Optimal tail design in species in which flight speed is important will be different from those in which maneuverability is more important than speed. Thus, birds that hunt aerial insects in open space are likely to have long narrow wings and large forked tails whereas birds hunting insects in habitats cluttered with vegetation are likely to have short broad wings and short tails.

How does all of this information apply to the evolution of the tails of male hummingbirds? First, long and/or elaborately designed tails impose an energetic cost based on the increased drag they produce. Tail length manipulation experiments in male Anna's hummingbirds, for example, indicate that both increasing and decreasing tail length increases the metabolic cost of flight by a modest 2–11 percent. Female preferences to mate with males with long tails, however, can override this cost. Such preferences are known to occur in a variety of birds. And once a preference for long or elaborate tails evolves, it can be self-reinforcing until the benefits of the preference are counteracted by its costs (as perhaps imposed by predators). Sexual selection via female choice is therefore very likely to be involved in the evolution of the elaborate tails found in some male hummingbirds.

The ancestral tail condition in hummingbirds is a rounded tail that is currently found in about half of all species, although the two basal clades of modern hummingbirds (Topazes and Hermits) do not have rounded tails. Males with unusually long or otherwise fancy or nonstandard tails are not concentrated in one particular hummingbird clade and have evolved independently many times. As a result, they are generally scattered throughout the family but are especially common in the Brilliant and Coquette clades (table 4). In Peru, a country with about 124 species of hummingbirds, notable examples of species with derived (non-ancestral) tails include many species of Hermits in which both males and females have long inner tail feathers; the swallow-tailed hummingbird (an Emerald); the fiery topaz (a Topaz) and black-bellied and wire-crested thorntails (Coquettes) in which males have very long inner tail feathers; and the marvelous spatuletail and booted racket-tail (both Brilliants) with teardrop-shaped tips on their outer tail feathers. It would be interesting to know under what environmental conditions these nontypical tails have evolved.

HOW HUMMINGBIRDS HARVEST FOOD

Long, narrow bills easily distinguish hummingbirds from swifts and other Strisores and signify their main method of feeding: probing into flowers to obtain nectar. Their highly specialized tongues are also essential for this kind of feeding. As I have mentioned, the evolution of hummingbird bills has been far less constrained than has the evolution of their flight machinery. Thus, hummingbird bills range in length from less than one centimeter (in thornbills, a Coquette) to more than ten centimeters (in the swordbill, a Brilliant). In shape they range from straight in many species to strongly decurved in Topazes and slightly decurved in Hermits and many others to slightly upcurved in the mountain avocetbill (a Coquette) (table 4). Average bill length in hummingbirds is about 19 mm and ranges from 8 mm to 119 mm.

Hummingbird tongues are critical for quickly acquiring nectar from the flowers they visit. Their tongues are about the same length as their bills but because they are attached to greatly elongated hyoid (throat) bones (as also seen in many woodpeckers), they can be extended far beyond the bill tip, greatly lengthening their food gathering reach. The tongues themselves are rigid and grooved, and their anterior half is split in two with fringed outer edges (figure 11). When these birds feed, the tongue is flattened and its two tips spread apart as they contact nectar. When this happens nectar is drawn onto the tongue and the tongue quickly resumes its preflattened shape as it is drawn back into the mouth where the nectar is squeezed out. In essence, then, the hummingbird tongue acts as an elastic micropump and not as a straw. In addition, specialized features of the tip of the bill of most hummingbirds, including forward-pointing serrations and small chambers beneath the tongue, have been hypothesized to act as "wringers" that further help to unload nectar into the mouth as the tongue is extended again during the next tongue lick. Results of recent very sophisticated videography of how hummingbirds unload nectar from their tongues supports this hypothesis.

The size and shape of hummingbird bills clearly influences the kinds of flowers they visit. Long-billed species tend to visit flowers exhibiting a greater variety of corolla lengths than short-billed species, and straight-billed species visit a greater variety of flowers than curve-billed species (figure 12). As a result, species with long straight bills are more generalized in their flower choices than species with either short straight bills or curved bills. Feeding efficiency—the amount of nectar birds can extract per flower visit—presumably lies behind these differences. Although small, short-billed species are mostly limited to flowers with short corollas when they visit tubular flowers, they can also successfully harvest nectar from many other kinds of non-tubular flowers that are mostly insect-pollinated. In a sense, then, the feeding niches of these hummingbirds (e.g., many Coquettes and Bees) are broader than their larger, straight-billed relatives that forage mostly at flowers with tubular corollas. Large species tend to visit insect-pollinated flowers only as a last resort when tubular flowers are scarce.

Although hummingbirds spend about 85–90 percent of their foraging time visiting flowers, they also need to eat insects to obtain the amino acids they require to build proteins. They capture insects, mostly flies, by two methods, and their bill shapes strongly affect these methods. Species with strongly curved bills (e.g., many Hermits) are hover-gleaners that pluck insects from spider webs and leaves in tropical forest understories. In contrast, straight-billed species often capture insects either by gleaning or by aerial hawking. In my yard, I often see straight-billed Anna's and Costa's hummers using both of these methods to capture tiny insects. The bills of straight-billed hummers have an unusual morphological adaptation that increases the size of their gape when they feed on aerial insects: their lower mandibles can flex downward, thereby increasing the width of their gape. Mandibular flexion is unknown to occur in other birds.

FIGURE 12 Four hummingbirds in action: A. booted racket-tail (*Ocreatus underwoodii*), B. green hermit (*Phaethornis guy*), C. Costa's hummingbird (*Calypte costae*), and D. broad-tailed hummingbird (*Selasphorus platycercus*). A comes from Costa Rica and B comes from Ecuador; C and D come from southern Arizona. Photo credits: Ted Fleming.

As in several other morphological features (e.g., plumage, body size, wing size), the size and shape of hummingbird bills are sexually dimorphic in many hummingbirds. General trends within species include (1) female bills are often longer than male bills; (2) the mating system (i.e., lek vs. non-lek systems; see below) affects sex-based bill length differences; lek mating males have longer bills than females; and (3) plumage dimorphism affects bill length dimorphism; sexual bill length differences decrease as degree of plumage dimorphism increases. These trends indicate that the evolution of sex-based differences in bill length in hummingbirds is complex and, in addition to its implications for feeding differences, involves both the form of the mating system and a species' plumage characteristics. In highly iridescent species, for example, males are particularly aggressive and tend to dominate access to rich flower resources regardless of bill length whereas bill length matters regarding access to flowers in less iridescent species. As a result, selection for sex differences in bill length should be stronger in less iridescent species than in iridescent species.

Two examples will illustrate this complexity. First, consider hermit hummingbirds, which are lek-maters (i.e., several males congregate in a small area and compete for the attention of females using visual and vocal displays), show relatively little sex-based size dimorphism, and have dull, monochromatic plumages. Based on a study of twenty-one members of this clade, the following trends have emerged: (1) females in many species have bills with greater curvature than males whereas males have longer bills than females; (2) the ancestral bill condition in this clade is females that have slightly curved bills; degree of female bill curvature increases in larger, more derived species; dimorphism in bill curvature has been lost in the sicklebills, in which both sexes have extremely curved bills; and (3) degree of dimorphism in bill curvature increases with bill length; thus, it is greater in larger species. Since it is known that bill shape influences flower choices in hummingbirds, it is likely that these sex-based differences in bill shape and size in Hermits have an ecological basis. As discussed in more detail in the next example, to reduce intraspecific competition for nectar, curve-billed females probably feed at a different set of flowers than males. Specifically, females likely feed at flowers with strongly curved corollas whereas males likely feed at flowers with straighter corollas. Finally, a new theme in discussions of sexual dimorphism in the bills of hermit hummingbirds has emerged with the discovery that males in the long-billed hermit, a common species that is distributed from southern Mexico to northwestern South America, have very sharp, daggerlike bill tips that are lacking in females. Lek-mating males in this species often fight among themselves and have been seen to stab each other in the throat. If females of this species choose mates based on their fighting ability, this would provide evidence for sexual selection for an effective weapon in males of this species. Such weapons are extremely rare in birds.

The second example involves the purple-throated carib, a member of the Mango clade that is widespread in the Lesser Antilles. Although males are 25 percent larger than females, the bills of females are 20 percent longer and 40 percent more curved than those of males—an extreme example of sexually based bill dimorphism. As a result, we might expect these bill differences to have a strong effect on flower choice in this species. Like many hermit hummingbirds, this species commonly visits flowers of *Heliconia* plants, a common group of monocot megaherbs in the understories of neotropical wet forests. As we might predict, members of the two sexes do visit different flowers that match their bill characteristics. Thus, male purple-throated caribs visit the relatively short, straight flowers of *Heliconia caribaea* while females visit the long, curved flowers of *H. bihai*. Careful behavioral observations have found that both males and females have higher feeding efficiencies when they visit their preferred flowers than when they attempt to visit flowers of the "wrong" species. In these islands in which the species diversity of plants and their pollinators is low compared with tropical mainland communities, the purple-throated carib hummingbird is acting as if it is two species because of the differences in the bills of males and females. We'll return to the topic of bill morphology and how it influences hummingbird foraging behavior again in the discussion of the structure of hummingbird communities.

Finally, although hummingbirds clearly are adapted for feeding at sugar-rich flowers, they also consume insects to obtain the protein and other micronutrients that they need, especially during times of nesting and migration. In fact, some ornithologists refer to them as insectivores that also visit flowers. Their ancestry as advanced members of the Strisores clearly has deep roots in insectivory as the primary feeding method. So what kinds of insects do hummingbirds eat and how do they obtain them? And how do their insect diets compare with those of swifts, their sister family?

As we know, hummingbirds are small birds that exhibit an impressive array of bill types. Thus, their size and bill morphology place significant constraints on the kinds of insects and other arthropods that they eat. As mentioned above, curve-billed hermits tend to glean spiders and insects from webs and leaves whereas straight-billed species are aerial hawkers of small flying insects. Hermits feed nearly exclusively in the forest understory whereas non-hermits feed both in the understory and in the forest canopy. Detailed analysis of the stomach contents of eleven species of hummingbirds in a lowland forest in northeastern Costa Rica indicated that spiders were by far the most common prey items in five species of hermits whereas wasps and flies were by far the most common prey items in six species of non-hermits.

Rather than sacrificing birds and examining their stomach contents to determine arthropod diets, as Gary Stiles did in the above Costa Rican study conducted in the early 1970s, researchers can now use DNA-based methods such as bar coding (i.e., the use of small species-specific segments of a particular mitochondrial gene) to identify animal and plant food items in fecal samples. An example of this approach comes from a study of the arthropods collected by female rufous hummingbirds to feed their nestlings on Vancouver Island, British Columbia. This noninvasive study used fecal pellets deposited under nests as sources of study material. Results indicated that females were capturing species from three arthropod classes, eight orders, forty-eight families, and eighty-seven genera with most of the prey being soft-bodied dipterans. We need to remember that this high prey diversity was documented at the northernmost extent of the range of hummingbirds in North America. It could be mind-boggling to consider the diversity of prey mamma hummers feed their babies in tropical habitats!

It is easy to visualize how curve-billed hummers are able to precisely pluck their prey from webs or leaf surfaces, but how do straight-billed species catch their prey? Do they snatch them from the air using their bills as mini forceps? Rather than plucking insects from the air with tweezers, however, hummingbirds are able to expand the base of their jaws both laterally and dorsoventrally and engulf their prey with an enlarged mouth, much as other aerial feeders such as swifts and swallows do. I recently watched an Anna's hummingbird do this and was amazed at how wide she could open her mouth as she captured small flies.

Since hummingbirds and swifts are sister families with very different lifestyles (table 3), it is of interest to compare their arthropod diets. How different are the animal diets of full-time aerial insectivores (swifts) from those of their smaller, nectar-feeding relatives? The answer to this question is somewhat surprising. For example, a study of the arthropods collected by three species of swifts in coastal Venezuela in food boluses that were being fed to their nestlings revealed that these birds were capturing a variety of spiders and insects from nine orders and 110 families. As in hummingbirds, small (2–3 mm) soft-bodied Hymenoptera and Diptera were most common, but their diets also included Hemiptera (true bugs), Homoptera (sucking bugs), Coleoptera (beetles), and Lepidoptera (butterflies and moths) that are not found in the diets of hummingbirds. In contrast, a temperate zone study based on nestling fecal samples collected from two species of swallows and one swift in southwestern Poland found that although these birds were capturing a similar array of insects, the diet of the swift was less diverse and contained smaller prey items than that of the swallows. The diet of the Polish swifts was dominated by hard-bodied arthropods such as beetles and ants. Similar results come from a long-term study of guano samples of chimney swifts in Ontario, Canada. Beetles and Hemiptera were the main dietary items in this study. Results of these studies suggest that the arthropod diets of tropical swifts are apparently more similar to those of hummingbirds than those of temperate zone swifts. Nonetheless, swifts likely capture somewhat larger and definitely harder prey items than hummingbirds.

WATER, WATER EVERYWHERE: THE WATER ECONOMY OF HUMMINGBIRDS

We've already reviewed the role of nectar, which is basically a liquid- and sugar-rich but protein-poor source of food, in the energy metabolism of hummingbirds, but the question remains, how do these birds manage their water economy and electrolyte balance? Given the nature of their diet, one important physiological problem that they face is getting rid of most of the water they ingest without losing valuable electrolytes. Most birds have kidneys that conserve electrolytes while producing concentrated urinary wastes, including water and uric acid, the nitrogen-containing waste produced by reptiles and birds. But hummingbird kidneys are not exceptional in this, and so they void lots of water as dilute urine as well as uric acid. They do manage to conserve some electrolytes in the cortex of their kidneys, but again they are not exceptional in this. As Carol Beuchat and colleagues have noted, hummingbirds are unique among birds in having kidneys that structurally resemble those of reptiles but whose rate of water flux resembles that of an amphibian. They also state (1990, p. 1059), "The simultaneous regulation of water and energy balance in hummingbirds consequently involves the complex integration of renal and intestinal functions and of these physiological processes with behavior and ecology." In the context of life in the universe, not to mention life on Earth, these birds are truly unique products of organic evolution, as I have repeatedly emphasized.

I can see and appreciate the importance of water balance in birds here in the desert every day. We have a bubbling fountain on our patio, and many birds and an occasional bobcat visit it. The birds include sparrows, finches, doves, hawks, and perhaps surprisingly, hummingbirds. Small songbirds and hummers both drink and bathe in it; the larger species just drink. Why do hummingbirds drink at our fountain when the food they eat is mostly water?

Calculations for an Anna's hummingbird weighing 4.5 g and living in southern California can give us a feeling for its rate of water intake from nectar and output via urine each day. Its normal intake is about 7.4 mL of nectar per day and increases to about 15 mL per day during periods of high activity. These values are the equivalent of about 1.6–3.3 times its mass and are much higher than expected for a bird its size. Its urine output is about 5 mL per day or 111 percent of its body mass. Again, at high levels of activity, it will excrete about three times its mass in urine per day—a value much higher than an equal-sized freshwater amphibian. Unlike most other birds, in hummingbirds most of the fluids they ingest pass quickly into and through the intestine to the cloaca (the avian equivalent of our rectum) without first being filtered and concentrated by the kidney, where electrolyte recovery occurs. Because of the reptilian nature of hummingbird kidneys, the fluid that does pass through them is dilute, not concentrated from the removal of electrolytes, and it reaches the cloaca as dilute urine. Nonetheless, electrolyte retention by the kidneys must be high enough for hummingbirds to remain in electrolyte balance under most conditions. Recent studies indicate that their kidneys do indeed play a more important role in electrolyte retention and urine production than previously thought. Nonetheless, unlike most birds and mammals, they really are living on the edge regarding their electrolyte balance.

Returning to my observations of hummingbirds drinking water at my fountain on warm days, they could be doing this to avoid becoming dehydrated. A combination of warm days (and nights), a small body size with its high surface to volume ratio, and a high body temperature and high metabolic rate means that their rate of water loss via passive evaporation from their skin and via panting is likely to be high. Ironically, whenever this occurs, hummingbirds can be in danger of becoming dehydrated, even when their food intake is mostly water. This danger is especially critical at night when hummers are not feeding. This could explain why they begin visiting our fountain shortly after sunrise on warm days.

An additional question concerning the unusual diet of hummingbirds might be: how do they avoid becoming diabetic? When they aren't feeding, their blood glucose levels are about 300 mg sugar per deciliter (dL) of blood (i.e., about three times higher than is normal in nondiabetic humans);

during feeding these levels increase over twofold to about 740 mg sugar per dL. Despite these high levels of blood sugar, hummingbirds do not display the usual symptoms (in humans) of glucose-enriched urine and a desire for excessive eating and drinking. High rates of glucose oxidation by flight muscles undoubtedly serve to help hummingbirds avoid the negative effects of hyperglycemia. Their rates of energy (sugar) intake are matched by equally high rates of energy use.

HOW DO HUMMINGBIRDS PERCEIVE THEIR WORLD?

As we know, most hummingbirds are very tiny birds. We've already learned that their small size strongly influences their metabolic rates and other aspects of their physiology. How does their size influence their sensory systems and how they perceive their world? How unusual are their brains and their sensory systems? Are they as unique in these features as they are in other aspects of their biology? And what is their *umwelt* (to quote Ed Yong)—their perceived world—like?

Let's begin this discussion by considering their brains. First, Haller's rule states that small members of any taxonomic group (e.g., hummingbirds) have larger brains and eyes relative to their size than larger members. This rule holds across a wide spectrum of vertebrates and invertebrates. Furthermore, since the brains of birds and mammals are known to require a substantial fraction (perhaps 20 percent) of their entire energy budgets for maintenance, this means that the brains of small species (i.e., < 4 g in hummingbirds) are more costly than those of large species. Small species have several options to accommodate their large brains. These include: (1) increasing the overall size of their skull; (2) changing the shape of their skull; and (3) reducing the degree of ossification of their skull. Regarding option (2), a spherical (round) skull can accommodate a larger brain than other shapes. Option (3) means that skulls of small species are less rigid and perhaps can accommodate relatively larger brains than the bonier skulls of large species. An analysis of the skulls of 96 species of hummingbirds supports options (2) and (3) but not option (1). It found that, compared with large species, small species did indeed have rounder and less ossified skulls and larger eyes but not larger skulls. Overall, however, all hummingbirds have larger brains relative to their size (up to 4.2 percent of their mass) than many other birds. This may result from their exceptional flight abilities or their need to track the changing locations of nectar sources in time and space. Reliance on spatial memory and vocal learning are two life history features that may require larger brains.

One distinctive brain-related characteristic of hummingbirds is that, along with parrots and advanced songbirds, they have the ability to learn specific songs; they exhibit vocal learning, a rare ability in birds. Working with Anna's hummingbird, this was first demonstrated experimentally by Luis Baptista and Karl Schuchmann in 1990. Hummingbirds are notable for the high diversity of their vocalizations (songs), and it is now known that they acquire them through imitation rather than by instinct. For example, male hermit hummingbirds within a lek learn each other's songs, and males in different leks sing somewhat different songs. Like parrots and songbirds, hummingbirds are known to have seven discrete structures called vocal nuclei in their forebrains that are active during singing. Locations of these nuclei in the brains of parrots, songbirds, and hummingbirds have been mapped using genetic transcription factors that are active when birds are singing. These vocal nuclei have been found in at least two hummingbird clades (Hermits [ancestral] and Emeralds [advanced]), suggesting that vocal learning may occur throughout the family. These nuclei are absent in the brains of birds that are only instinctual singers.

In addition to relatively large brains, hummingbirds have relatively large eyes. In terms of absolute size, however, their eyes are still very small, and this limits their visual acuity. When I watch hummingbirds sitting quietly between feeding bouts in my yard, they are always moving their heads and looking all around. As a result, they are very quick to spot another hummer as it approaches flowers or a feeder. This might lead one to conclude that they have very keen vision, but this not correct. Because of the small absolute size of their eyes, their visual acuity is not exceptional. Results of laboratory tests indicate that species such as Anna's hummingbird have 20/100 vision compared with hawks and owls whose much larger eyes give them 20/4 to 20/9 vision.

Like most birds, hummingbirds have tetrachromatic vision, which means that their retinas have cones (color-sensitive cells) that are "tuned" to four different regions of the visible electromagnetic spectrum, which ranges from about 400 to 700 nanometers: very short (UV), short (blue), medium (green), and long (red) wave lengths. Peak sensitivity of their cones occurs at 560 nanometers with a secondary peak in the UV. We only have trichromatic vision and cannot see UV wavelengths. As a result, the visual world of hummingbirds is different from ours. It is much more colorful than ours. It's as if we're color-blind whereas birds are not. Given their ability to see short-wave

(UV) colors and in conjunction with the oil droplets in their cone cells, birds can see color combinations such as ultraviolet-green and ultraviolet-yellow that we cannot even imagine. Hummingbirds use this expanded color palette, in part, for choosing the flowers that they visit. In addition, many female birds use this color palette to discriminate among potential mates based on slight differences in the (UV) colors of their plumage. In other birds, ultraviolet reflectance is important in choosing which nestlings to preferentially feed, which fruits to eat, and, in vole-eating kestrels, where to search based on concentrations of rodent urine and feces, which reflect ultraviolet wavelengths.

Several adaptations give hummingbirds a unique visual system. First, their eyes are laterally placed, which gives them a very broad field of view but a narrow binocular field of view. Second, like all birds, their retinas have one centrally placed indentation called a fovea and a flat lateral area called the area temporalis, both of which contain high concentrations of neural ganglion cells for enhancing visual acuity. The central fovea is laterally focused, which gives them excellent vision for distant objects in their lateral field of view. The area temporalis is forward focused and assists with central and binocular vision. In contrast, the retinas of mammals have only one forward-facing fovea. Third, they possess a specialized visual response that focuses on objects immediately in front of them that allows them to remain stationary at moving flowers and to precisely track other moving targets such as birds and flying insects. To do this they have a much larger area of their midbrain containing neurons that encode for optic movement than other birds, including swifts. And in contrast to animals such as bees, flies, and parakeets, this neural region is more sensitive to movements of objects in front of a hummingbird than at its sides, which allows hummingbirds to pursue fast-moving objects in both open and cluttered environments. In addition, recent research indicates that the visual system of hummingbirds is unique among birds in two other respects. First, it only reacts to fast-moving objects rather than to both fast- and slow-moving objects as in other birds and mammals; as a result these birds can respond very rapidly to slight changes in the location of items such as swaying flowers or fleeing competitors or mates in front of them. And second, their visual system can respond to images within a 360-degree area around them rather than only 180 degrees in the horizontal plane as occurs in other birds; this ability is thought to increase a hummingbird's visual stability during hovering.

Finally, an analysis of the effect of body size (mass) on visual spatial and temporal resolution has indicated that spatial resolution is positively correlated with mass whereas temporal resolution is negatively correlated with mass. This means that very small birds such as hummingbirds have relatively low visual spatial resolution because of their small eyes but have relatively high visual temporal resolution because of their high reaction speeds. The few nocturnal birds (owls) that have been tested have lower visual spatial and temporal sensitivity for their size compared with diurnal birds. This difference is interesting because of the evolutionary history of hummingbirds, whose ancestors were nocturnal. It is likely that many features of the visual systems of early nocturnal Strisores such as their color vision (or lack thereof) were very different from those of their diurnal relatives. As we've previously seen, a diurnal lifestyle has favored the evolution of a whole host of different adaptations in hummingbirds compared with those of their nocturnal ancestors.

HEARING AND VOCALIZATIONS

Birds are perhaps the most vocal of all vertebrates, and the sounds they produce are therefore critical elements of their sensory worlds. Birds vocalize to attract mates, defend territories, alert others about predators, and to simply keep in contact with each other. In doing this they produce two kinds of sounds: songs and calls. Songs often have a complex structure and are produced mainly by males during the breeding season; calls are much simpler and are produced by both sexes throughout the year. Although oscine (advanced) passerines are champion songsters, hummingbirds are no slouches either regarding their ability to sing, though we tend to think that passerine songs are much prettier to our ears, on average, than hummingbird songs. But who knows whether hummingbirds consider their songs to be just as beautiful as the melodious songs produced by house finches? After briefly reviewing how birds produce, hear, and process their vocalizations, I'll discuss vocal adaptations in hummingbirds. In box 1, I compare the songs of an Anna's hummingbird with that of a common songbird, the house finch.

All vocalizations require a mechanism for producing them and a mechanism for hearing and interpreting them. Sound production involves a syrinx, a unique structure found only in birds. In most birds, including swifts and nightbirds, the syrinx is located deep in their body at the junction of the trachea and the first two major bronchi. It is composed of thin syringeal membranes in the air passageway, supporting elements of cartilage or bone, muscles that attach to the membranes and supporting elements, and nerves that control contraction of the syringeal muscles. In hummingbirds, however, the syrinx is located

Box 1. What's in a bird song?

As we all know, one of the most distinctive features of birds is their vocal prowess. Most of the passerine birds around us here in the Temperate Zone are dedicated vocalists to the extent that many birdwatchers identify species by their songs first before they even see them. Experienced tropical ornithologists such as the late Ted Parker are said to have been able to identify hundreds of species of birds in hyper-diverse tropical bird communities solely by the songs and sounds they produce.

It's not difficult to understand why many passerine birds, with their specialized vocal organ (the syrinx), are called songbirds. Many of them produce very beautiful songs. Just think of the flutelike songs of wood thrushes in the hardwood forests of the eastern United States; or the cardinal-like songs of Carolina wrens in the southeastern United States; or the astonishing array of melodies that northern mockingbirds routinely sing throughout southern North America. To be sure, some passerines such as ravens and crows aren't very musical, but, taxonomically, they still are songbirds.

Given the rich array of songs produced by passerines, it may seem strange to consider hummingbirds, also equipped with an independently evolved specialized syrinx, to be songbirds. But they certainly are, particularly during the breeding season when the vocal abilities of all male birds are at their best. It's just that hummingbird songs are clearly different from those of many passerines. Instead of producing songs such as those of the starling and canary that Mozart loved and incorporated into his music, hummingbird songs are very buzzy and insect-like. But that's because their song structure is very different from that of more melodious songs.

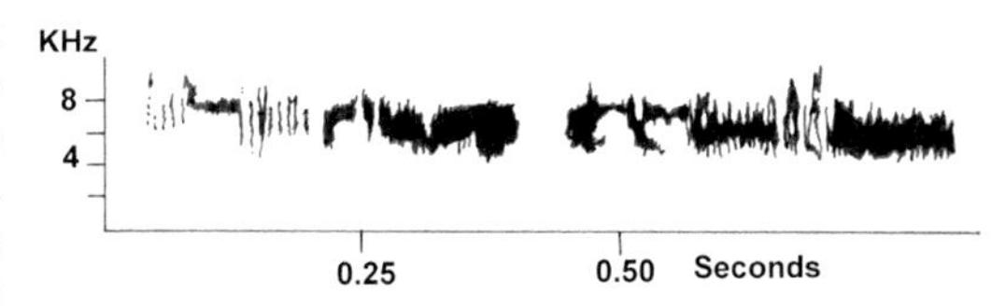

An example of two songs produced by a male Anna's hummingbird. The frequency marks on the y-axis are 0, 4, and 8 kHz, and the duration of each song (x-axis) in less than 0.5 second. Credit: Baptista and Schuchmann (1990).

To see this, let's compare a song of Anna's hummingbird with that of the house finch, two birds that sing in my backyard. The standard way of visualizing them is by a sonogram, which is a plot of the structure of sounds with their frequencies in kilohertz (kHz) depicted on the y-axis and time on the x-axis. From this, it is easy to see that these two songs look (and definitely sound) much different. Both songs cover the typical frequency range of bird songs (i.e., 2–8 kHz, which is well within our hearing range). But that's where the similarity ends. The hummingbird song is short and very dense (i.e., frequency-rich) and "buzzy" like that of an insect. It lacks any semblance of a melody with clear notes or pure tones. In contrast, the house finch song is longer and contains a series of relatively pure notes that sweep and swirl through a series of frequencies with a dense "buzz" at the end. This structure produces a much more musical sound.

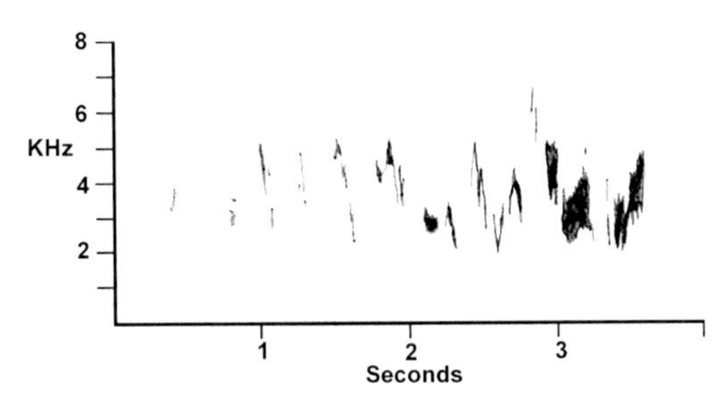

An example of a song produced by a male house finch. Time on the x-axis is seconds. Credit: Tracy and Baker (1999).

higher in the throat outside the thorax, perhaps because their enlarged heart occupies much of the chest cavity. Also, details of its syringeal musculature differ from those of oscine songbirds, indicating that the hummingbird syrinx has evolved independently from that of passerine songbirds.

Sound reception in birds occurs in their ears, whose bony structure is basically similar to that of mammals, except, as a reflection of their reptilian ancestry, sound is transmitted from the tympanic membrane to the inner ear via a single bone—the columella—rather than by a set of three ear bones. The typical hearing range of birds is 2–8 kHz, and this is also true for many species of hummingbirds. For comparison, our hearing range is much broader: 2 Hz–20 kHz. In birds, sounds with frequencies greater than 8 kHz are usually called *ultrasound*; in humans, *ultrasound* refers to frequencies greater than 20 kHz. It is now known that some hummingbirds produce ultrasound in the range of 15–20 kHz. These species are found in at least five of the nine hummingbird clades (Topazes, Brilliants, Coquettes, Mountain Gems, and Bees; table 4). Depending on species, these high-frequency sounds are used for territorial defense or courtship. For example, during courtship male Ecuadorian hillstars produce high-frequency sounds centered on about 13 kHz that females can hear and respond to. In southern Arizona, males of Costa's hummingbird, but not their sister species Anna's hummingbird, also use these sounds in their courtship displays.

BRAINS AND SOUND RECEPTION

The neural aspects of sound reception and production have been worked out in great detail in advanced (oscine) songbirds. The general system in these and other song-producing birds involve three basic neural pathways: the input pathway, the sound production pathway, and, in song learners, the anterior or forebrain regions involved in song learning. The input pathway brings sounds from the ears to neural centers located in posterior regions of the forebrain or cerebrum; the sound production pathway is also located in this part of the cerebrum where it connects to the input centers and then sends nerve impulses to the syringeal muscles, larynx, and respiratory muscles for song production; and the anterior region of the cerebrum contains neural centers for evaluating and producing learned songs. Using sophisticated methods based on mapping estrogen receptors and genetic techniques, it is known that the brains of hummingbirds contain neural nuclei that are analogous to the same brain centers in song-producing passerines. As a result, they have independently evolved the same neural circuitry to become accomplished singers and song learners.

MORE ON VOCAL LEARNING

We know that hummingbirds are unique among birds in many ways (table 3), and one of these ways is that they are vocal learners. Like many well-studied oscines, juvenile hummers have a sensitive period during which they are receptive to learning songs. This is interesting because, unlike oscine songbirds, male hummers are not involved in raising their young who must therefore learn songs without close contact with their fathers. Nonetheless, young hummers must be keenly aware of their acoustic environment in and after they leave the nest. And this awareness must involve ignoring most of their acoustic environment in favor of distinctive hummingbird sounds that they learn. They presumably learn their songs from nearby males. This raises the question, however, of whether they are exclusively tuned into songs of their own species to the exclusion of the songs of other male hummers. For example, males of both Costa's and Anna's hummingbirds sing in our yard in the spring. Do juvenile male Anna's learn only those of the "correct" species and vice versa for Costa's? The fact that these two species are known to hybridize where their geographic ranges overlap suggests that these events may occur where males are singing the "wrong" song. Laboratory studies have shown that, like oscine songbirds, young hummers learn their songs more readily from "live" tutors than from tape-recorded songs. However, unlike most other avian song learners, hummers such as hermits can also learn new songs when they join a lek as adults. Thus, in hummingbirds, it seems that you can teach an old bird new tricks. Finally, the ability to learn songs means that geographic dialects are likely to be as common in hummingbirds as they are in many oscine songbirds. This certainly occurs in Anna's hummers in the western United States.

PUTTING IT ALL TOGETHER, PART 1: VOICES, TAILS, AND WINGS AS COMMUNICATION "INSTRUMENTS"

We've seen that two senses—visual and auditory—are involved in the evolution of courtship behavior in hummingbirds. Females are often attracted to brightly colored males and, sometimes, to males with unusual plumages (e.g., fancy tails or crests). In addition, sounds produced by singing males, and in Bees, sounds produced by their wing and tail feathers are important aspects of courtship behavior. Hummingbird researchers are now asking whether learning that occurs in both visual and vocal displays has coevolved together in these birds. Since song production and visual displays both involve complex motor patterns, it seems reasonable to propose that they might evolve (coevolve) in parallel. Evidence supporting this hypothesis comes from work with the long-billed hermit hummingbird in the tropical lowlands of Costa Rica. Like other hermits, this is a lek-mating species in which groups of males congregate in small areas and use both vocal and visual displays to attract females. Since hermits are sexually monomorphic and dull colored, their visual displays include a series of stereotyped movements (e.g., bill posturing and short flights to change perches) and hence are conspicuous aspects of their courtship behavior. Results of a detailed analysis of variation in vocal and visual displays within leks and between leks in three different areas indicated that, as predicted, small-scale variation in both song and visual displays exists in the absence of any genetic differences within and between leks and areas. This variation is consistent with the idea that learning of the elements of visual displays accompanies learning of different song types on a small spatial scale. Therefore, it appears that song production and visual displays are learned behaviors that have likely coevolved together in this species.

Further support for the idea that vocal and acoustic elements of courtship displays in hummingbirds sometimes coevolve together comes from a study of variation in the courtship behavior of Bee hummingbirds, an advanced clade of about thirty-seven species. As we've already seen, males of these hummers are notable for producing sounds by their tail feathers during courtship flights. Additional sounds are produced by wing trills and vocalizations. A detailed analysis of the occurrence of these three kinds of sounds conducted within a phylogenetic framework has revealed that during courtship flights (1) vocalizations and wing trills are negatively correlated; wing trills replace vocalizations as a means of sound production during these flights in some species, probably because they are redundant in the quality of the sounds they produce; and (2) there have been three instances of convergence (coevolution) between the characteristics of sounds birds produce vocally and via their tails. In this clade, the occurrence of courtship dives probably evolved before tail-produced sounds and vocalizations were added to the dives. Sexual selection for different tail sounds during dives by females then resulted in changes in the shapes of tail feathers. Overall, the evolution of acoustic-based courtship displays in these birds has probably occurred rapidly—over a period of five million years or less. And it has resulted in a high diversity of different display characteristics. For the males in this clade, no single answer has emerged to the age-old problem of how one can be maximally attractive to females. Instead, there have been multiple answers to this question.

PUTTING IT ALL TOGETHER, PART 2: THE COGNITIVE ABILITIES OF HUMMINGBIRDS

Like all mobile organisms, hummingbirds are faced with a myriad of decisions every day, including where to feed, what to eat, how to attract mates, and how to avoid competitors and predators. Dealing with these kinds of issues in an ever-changing environment involves their cognitive ability, defined as the ability to recognize and deal with problems using thinking and learning skills. Since hummingbirds have large brains relative to their size and are skilled song learners, we might expect them to have well-developed cognitive abilities for dealing with life's challenges. Because the study of their cognitive abilities in the field is a relatively new research area, little literature exists yet for hummingbirds. But the following studies should give us some insight into this important aspect of hummingbird biology.

The first study deals with a familiar species—the lek-mating long-billed hermit hummingbird in Costa Rica. In that study researchers sought to determine whether good spatial memory concerning the location of food resources, a known aspect of cognitive ability, was correlated with the ability of males to obtain and hold a display territory within a lek. Good spatial memory should result in efficient foraging that gives males more time to occupy their display territories. Duration of territory occupancy is especially important because females visit each lek for mating purposes only about once a day. Spatial memory for a series of individually marked territory-holding and nonterritory-holding ("floaters") males was assessed in the field using arrays of three hummingbird feeders placed near several leks; one feeder contained 25 percent sugar water and the other two contained only water. After birds had initially tested the feeders and discovered which feeder contained sugar, they were scored on their next return visit as to whether they visited the "correct" feeder first or not. After one trial was completed, the positions of the feeders were changed, and males were reassessed again for a total of ten trials per individual. Each individual's spatial memory was scored as percent correct decisions in the ten trials. In addition, a male's size (mass), bill length, and the consistency with which it produced temporally stable display songs were also examined as possible factors predicting a male's ability to hold a territory on a lek. Results indicated that two of these factors—spatial memory and body mass—were the best predictors of a male's ability to hold a territory. Of these two factors, spatial memory—the cognitive factor—scored highest as a predictor of a male's territory-holding ability. Consistency of song production was also positively correlated with a male's spatial memory score. Although male mating success was not determined in this study, previous work has shown that the amount of time a male spends displaying on its territory increases its attractiveness to females. Thus, cognitive ability as expressed as spatial memory and song consistency appears to be important for a male's reproductive success in this species.

The second species is the green-backed firecrown (a Coquette) from the lowlands of Chile and western Argentina; unlike the long-billed hermit it is a territorialist and not a lek mater. Its spatial memory was tested in two experimental designs. The first one was the same one used for the long-billed hermit in Costa Rica; males were scored regarding their ability to correctly return to the one feeder containing sugar water in a series of feeders that contained only water. The second one was a slight modification. The correct feeder contained 25 percent sugar and other feeders contained less sugar. In both designs, males had high scores for choosing the correct or best feeder, confirming that their cognitive ability is high.

In addition to having good spatial memory, hummingbirds have been shown experimentally to be able to anticipate the nectar renewal rates of flowers and to match their visitation patterns to coincide with these renewal rates. Experiments with green-backed firecrowns indicate that they can integrate information about the location of good feeding sites with nectar quality and nectar renewal rates. Other experiments have shown that hummingbirds use visual cues such as flower color or a distinctive marking, in addition to spatial location, nectar quality, and nectar renewal rate, to influence their feeding decisions. Overall, these results indicate that these birds have impressive cognitive abilities. Environmental variation in the flower resource base is probably a key factor in selecting for these cognitive abilities. Supporting this, it is known that migratory hummers actually have better performances in cognitive tests than those that live in less variable environments and are nonmigratory.

ADDITIONAL HUMMINGBIRD SENSES: SMELL AND TASTE

Unlike mammals, most birds have a poorly developed sense of smell (although this view is changing based on recent research), and this is generally true of hummingbirds. A study of the relative size of the olfactory bulbs of five non-hermit hummingbirds found that their bulbs are the second smallest among twenty-five orders of birds. Reflecting this, most hummingbird flowers produce no scent, as I discuss in more detail later. The results of an experiment with captive *Amazilia* hummingbirds (an Emerald) indicated that these birds did not discriminate between experimentally scented and unscented flowers. But the results of at least one experimental olfactory-based study indicates that the white-vented violet-ear (a Mango) can discriminate between sugar vs. salt solutions based on natural nonfloral fragrances associated with them. Compared with the effect of a red or yellow color associated with the feeder containing each solution rather than odor, however, this species responded more strongly to color cues, supporting the idea that hummers use visual information rather than odor for their floral choices. One situation in which Bee hummingbirds have been shown to use olfactory information to make floral choices is when the smells produced by ants and wasps, but not honeybees, occur in floral nectar. They avoid flowers containing ant and wasp scents, both of which are competitors for nectar, but not those containing honeybee (an exotic species) scents. Subsequent research, however, has shown that taste, rather than smell, is likely involved in this discrimination.

Scented flowers are common in insect-pollinated flowers, the ancestral condition in the evolution of hummingbird pollination from insect pollination. Detailed genetic and biochemical research with moth- and hummingbird-pollinated *Petunia* flowers has found that a specific gene, designated CNL, which is involved in initiating floral scent production in moth flowers, was inactivated early in the transition from moth to hummingbird pollination. This inactivation apparently preceded the evolution of flower color from white to red in hummingbird-pollinated *Petunia*s. It is likely that a preference for unscented but red-colored flowers by hummingbirds lies behind this transition. Incidentally, it has been suggested that the lack of scent in most hummingbird-pollinated flowers has evolved to reduce their attractiveness to bees, thereby eliminating an important group of competitors for floral nectar.

Unlike their sense of smell, most birds, including hummingbirds, have a well-developed sense of taste. Their taste buds are located at the back of their tongue and the bottom of their throat. In the case of hummingbirds, a good sense of taste is important because of the wide variation in the sugar content and other chemical composition of nectars in the flowers they visit. Because of their high energy requirements, selection should favor the evolution of excellent discrimination abilities for choosing the flowers they visit. Ancestrally, the two taste receptor genes in vertebrates responded only to amino acids found in proteins and not to sugars. In songbirds and hummingbirds, however, these taste receptors have converged via gene duplication and refunctioning to become responsive to both sugars and amino acids. Survey of the response of taste receptors of hummingbirds from several clades indicates that the shift from sensitivity to amino acids to sugars occurred at the base of hummingbird radiation, after they had split from swifts. Their receptors still retain some sensitivity to amino acids, which varies among hummingbird species, but they are much more responsive to sugars, especially to sucrose, which is the most common sugar in flower nectars. Since certain amino acids interact synergistically with sucrose to affect hummingbird nectar choice, it is not surprising that some flower nectars still contain small amounts of amino acids. In addition, it is now known that some hummingbirds (e.g., Anna's) are also not averse to the taste of bitter compounds found in some floral nectars that repel swifts and nighthawks. This difference reflects changes in the two additional receptor genes that respond to bitter substances after hummingbirds split from their insectivorous ancestors. The ability of some hummingbirds to tolerate bitter-tasting substances makes it possible for plants to limit the attractiveness of their flowers to other pollinators by producing nectar that contains bitter substances.

HUMMINGBIRD FORAGING BEHAVIOR: OPTIMAL OR NOT?

In a sense, hummingbirds are little foraging machines. Their high energy requirements demand that they forage continuously throughout the day. Field observations of migrant hummers in western North America, for example, indicate that these birds visit flowers to feed every fifteen to twenty minutes all day during which they imbibe fifty to two hundred microliters of nectar per flower visit. They therefore ingest fifteen to twenty milliliters of nectar containing 20–25 percent sucrose every day—an amount of sugar equivalent to their body weight. Given the importance of obtaining enough energy each day, especially when birds are building up fat reserves during migration, we expect them to be experts at maximizing their foraging efficiency, which raises the question, Are they "optimal" foragers in some sense?

The concept of optimal foraging is concerned with identifying an animal's traits that are involved in maximizing an animal's rate of energy acquisition.

It became a popular topic in behavioral ecology beginning in the mid-1960s. Because hummingbirds are easy to observe in the field and their floral resources are easy to measure, they became popular subjects for optimal foraging studies, beginning in the early 1970s. As a result, an extensive literature on this topic now exists. Before reviewing some of this literature, I will first describe the basic foraging strategies used by hummingbirds to harvest floral nectar.

Hummingbirds use two general methods when foraging: traplining and feeding in territories that they defend. Traplining involves visiting widely spaced flowers in a regular circuit. This method occurs most commonly in Hermits and does not involve territorial defense of flowers. Male hermits only defend display territories on their leks. In contrast, feeding in many other hummers involves some kind of territorial behavior in which adult males (usually) defend a patch of flowers against intrusions by other birds of the same or different species. Females and juveniles are usually behaviorally subordinate to males and often must resort to sneaking into territories whenever males are away or are otherwise occupied. Also, within communities of hummingbirds, dominance hierarchies often exist in which some species are dominant over others regarding access to flowers. In subordinate species, both males and females become "sneakers" to gain access to flowers within the feeding territories of dominant species.

In 1978, Pete Feinsinger and Rob Colwell elaborated on this basic dichotomy by classifying hummingbird foraging strategies into five different types. These include high-reward (nectar) trapliners, low-reward trapliners, territorialists, territory parasites or "marauders," and generalists. Trapliners thus can be differentiated based on the kinds of flowers they visit. High-reward species, mostly Hermits, are larger and have curved bills compared with the smaller low-reward species with straight bills (e.g., Emeralds) (table 4). Territorialists tend to be medium sized and short billed (e.g., Brilliants and Mountain Gems). The "marauders" (e.g., Topazes and Mangoes) are often large and can easily invade territories held by smaller birds. And generalists tend to be medium sized and have small-to-medium bills (e.g., Brilliants, Coquettes, Bees, and Emeralds). They usually visit widely dispersed rather than clumped flowers and insect-pollinated kinds; they also sometimes sneak into territories to "steal" nectar. These different strategies reflect differences in their flowers (their size, shape, nectar content, and plant distribution pattern) and differences in hummingbirds (body size, bill size and shape, and the cost of hovering as determined by a bird's size and wing shape). Given that most tropical plant communities contain a complex mosaic of flower types, it is obvious that they are likely to be able to support many kinds of hummingbirds. We'll revisit this topic in more detail below.

Returning to the topic of optimal foraging, in 1971 Thomas Schoener published a comprehensive review of feeding strategies that discussed its theoretical issues and set the stage for many empirical studies. In his terminology, hummingbirds are Type II "predators" whose foraging strategy includes the time and energy involved in searching for and extracting energy from flowers. In this scheme, time and energy are key elements and from these two general ideas emerged. Optimal foragers could either be "time minimizers" or "energy maximizers." Time minimizers quickly obtain the energy they need so that they can devote time (and energy) to other biological needs, for example, attracting mates, interacting with competitors, and avoiding predators. To do this, they might not be particularly picky feeders in order to reduce their overall foraging time and are thus likely to be feeding generalists. In contrast, energy maximizers are likely to be picky feeders that carefully choose their food based on its energy content (or net reward). In their food choice, potential food items would be ranked by their energy content with high-ranking items being sought preferentially, even if these items take more time to find. Energy maximizers are therefore likely to be feeding specialists. Applying these concepts to hummingbirds, during the breeding season males are more likely to be time minimizers whereas females are more likely to be energy maximizers. Individuals of both sexes are likely to be energy maximizers whenever they migrate.

Does the foraging behavior of hummingbirds, both within and between species, conform to these expectations? As we've already seen, whether hummingbirds are likely to be feeding specialists or generalists strongly depends on the shape and size of their bills. Species with curved bills (e.g., Topazes and Hermits) are more likely to be feeding specialists than species with straight bills (e.g., Brilliants, Mountain Gems, and some Emeralds) (table 4). And species with long bills are more likely to be feeding generalists than those with short bills because they can effectively obtain nectar from both long- and short-corolla flowers; long-corolla flowers are generally not available to short-billed hummers. One feeding advantage that small- or medium-sized species with short bills (e.g., certain Coquettes, Bees, and Emeralds) have over other species is that they can effectively harvest nectar from flowers whose main pollinators are insects. Data based on the feeding records of many species of hummingbirds and published by Rodríguez-Flores and colleagues in 2019 support these trends.

The following examples provide an overview of the variety of feeding methods found in hummingbirds. First, consider the long-billed hermit, which, like other Hermits, is a lek-mating and traplining species. It typically develops a daily circuit that can be a kilometer or more in length in which it visits and revisits clumps of nectar-rich flowers produced by species of megaherbs of the genus *Heliconia*, among other kinds. It does not defend patches of these flowers, which are therefore available to potentially competing hummers. Major foraging concerns this bird has thus include knowing the nectar renewal schedules of the flowers they visit (we've already seen that hummingbirds have the cognitive ability to do this) and dealing with uncertainty caused by other birds visiting and depleting their flowers. Timing their revisits to flowers is a kind of classical operant conditioning system in which the bird's feeding behavior sets the reinforcement schedule.

Working in a lowland tropical forest in southwestern Costa Rica, Frank Gill and his assistants documented the foraging behavior of long-billed hummer males from one lek at their natural flowers and at hummingbird feeders placed at different distances from the lek. Nectar (sugar water) amounts and refill rates of the feeders were experimentally manipulated. They found that males tended to visit flowers to which they had sole access. Males from the same lek distributed themselves among available flowers so that there was little or no competition for nectar among them. When there was competition, birds visited flowers more often than when there was no competition in an effort to maximize their consumption of a particular flower's nectar. Under different renewal schedules and amounts of nectar at feeders, the birds' return schedules matched expected renewal rates and amounts of nectar. For instance, as they did for their flowers, they revisited feeders containing lots of nectar and high renewal rates much faster than they did at flowers or feeders that contained low amounts of nectar and low renewal rates. This suggests that they have a finely tuned sense of the nectar production schedules of their food plants and adjust their traplining schedules accordingly.

Next, let's consider the foraging behavior of a migratory territorialist, the rufous hummingbird (a Bee) of western North America, which overwinters in Mexico and Central America and breeds from Oregon to Alaska. In returning to Mexico in the early fall, it travels through the mountain meadows of the Sierra Nevada and Sierra Madre. During its southward migration, it stops for up to two weeks in the mountains to replenish its fat stores. In the study area of Mark Hixon and colleagues in eastern California, its principal food came from flowers of a species of Indian paintbrush. In this area, as I described earlier (page 46), migrant birds defend discrete territories and spend about 20 percent of the day foraging, 3 percent defending their territories against other hummers, and 77 percent simply sitting during which they empty their crops and ultimately deliver digested sugar to their flight muscles. At first glance, these birds thus appear to be time minimizers and not energy maximizers as predicted above. But they actually are energy maximizers because their rest time should be considered to be "food handling and processing time" and not simply "quiet time." Given their small size and limited crop volume, they simply cannot continue to ingest nectar once their crops are full. They need to take a break to empty their crops and replenish their energy supply while they rest.

In addition to determining their time and energy budgets, Hixon and colleagues experimentally manipulated the territory sizes of three birds by covering half the flowers in each territory for one day. Birds responded quickly to this manipulation by doubling the size of the area they defended, by increasing their amount of foraging time and the length of each foraging bout, and by decreasing their amount of resting time; the time they spent defending their territories did not change, probably because small territories are less attractive to intruders than large territories. When the flowers were uncovered and the original territory size restored, the time budgets of these birds quickly returned to premanipulation levels—an impressive demonstration of how responsive these birds are to changes in the amount of energy available in their environment. Overall, the researchers concluded that these migrant hummers are indeed energy maximizers during this critical phase of their annual life cycle.

Finally, we've seen the impressive cognitive abilities of green-backed firecrown hummingbirds (a Coquette), a territorialist and migratory hummingbird living in temperate habitats in southern South America. Further experimental work with this species indicates that, like the long-billed hermit, it can remember and adjust its foraging behavior in response to both nectar renewal rates and the sugar content of artificial feeders. This implies that it, and probably many other species of hummingbirds, has evolved to effectively fulfill its energy needs in the face of temporal and spatial variation in its natural resources, competitive pressures from other hummingbirds, and the energy demands associated with migration. Like most other animals on

Earth, hummingbirds live in complex worlds filled with many challenges to their overall fitness. Some of these challenges, including those dealing with the high energy demands resulting from its size-based metabolism, are especially acute, and the fact that these birds can thrive in a wide range of physical and biological environments is a testament to how adaptable they have been during their evolution.

MIGRATION: HUMMINGBIRDS ON THE MOVE

As I discuss in more detail below, the major resource base of hummingbirds—flowers and their nectar—is highly variable in space and time so that the suitability of their habitats to support these birds is constantly changing in many parts of their geographic ranges. As a result, many hummingbirds change habitats. Some move only short distances while others move up and down mountains. Still others make substantial intercontinental movements, for example from Central America and Mexico into the United States and Canada seasonally. In their review of these movements, Rappole and Schuchmann indicated that at least eighty-seven species are known to undergo altitudinal migrations, forty-two species undergo latitudinal migrations of up to one thousand kilometers, and an additional twenty-nine species undergo long-distance migrations of over one thousand kilometers. Reflecting the ubiquity of either short- or long-distance movements, in 1978 Feinsinger and Colwell pointed out that most hummingbird communities contain two kinds of species: principals and secondaries. Principals can be considered to be the "core" or resident species that remain within particular habitats year-round. Their feeding behavior likely determines the resources available to immigrants into these communities. Secondaries include species that move among neighboring habitats depending on flower availability and true migrants that travel longer distances, either along altitudinal or latitudinal pathways. These species often include high-reward trapliners (short-distance migrants) and opportunistic generalists (long-distance migrants).

In 1992 Doug Levey and Gary Stiles reviewed the migratory status of birds living in Costa Rica's wet Atlantic side whose avifauna these researchers know very well. Compared with insect-eating birds, fruit eaters and nectar feeders were much more likely to be either altitudinal or latitudinal migrants because of the greater seasonality of their food resources. Among frugivores and nectarivores, the latter group contains a proportionately greater number of altitudinal migrants than the former group. Overall, about two-thirds of the species of Atlantic zone Costa Rican hummingbirds undergo local or altitudinal movements. Hummers feeding in forest canopies are most likely to be altitudinal migrants whereas species feeding in second growth or along forest edges typically undergo shorter between-habitat movements. On the drier Pacific side of Costa Rica, nearly all species of hummingbirds living in and around tropical dry forests undergo seasonal movements.

One intriguing aspect of the migratory biology of a least some birds is that migratory species tend to have relatively smaller brains than their nonmigratory relatives. A detailed analysis of passerine songbirds, for example, has revealed this trend, both within the entire order and in specific families. Why natural selection would favor the evolution of relatively smaller brains in migrant species is not yet clear but may involve the competing demands of the high metabolic cost of maintaining a large brain versus the high metabolic costs involved in migrating, especially in long-distance migrants. This kind of analysis has not yet been applied to hummingbirds, so we don't know whether migrant hummers are smaller brained than nonmigrants. If they are, this might have important implications for differences in the cognitive abilities of migrants versus nonmigrants.

In summary, the flower-visiting habits of hummingbirds have resulted in the evolution of a very mobile lifestyle in many species. As an extreme example, consider the rufous hummingbird, a northern Bee species that winters in central Mexico and breeds in northwestern North America as far north as southern Alaska. The hummingbird biologist Bill Calder once claimed that for its size (3–4 g, total length about 9.5 cm), this species may hold the record for one-way migration distance—forty-nine million body lengths. And since individuals can live seven or eight years, think of all the wear and tear they place on their bodies and how much oxygen and nectar they consume during their lifetimes. It's pretty amazing, right?

HOW HUMMINGBIRDS MAKE BABIES

Since it takes two to tango, hummingbirds have to be a bit social at some time during their lives. But no one would consider these birds to be social. They certainly don't roost in large colonies or travel together in flocks like many other birds, including some swifts, do. Instead, most species live quite solitary lives. Robert Bleiweiss has reviewed the "social" systems of hummingbirds and recognizes four different spatial configurations based on aggregation patterns of the sexes. These include (1) both males and females dispersed; (2) males aggregated in leks, females dispersed; (3) males dispersed, females in small colonies; and (4) males aggregated in leks, females in small colonies. Of these, pattern (1) is probably most common as exemplified by all of the North American species. As I've indicated, lek mating is particularly common in Hermits but also occurs in a few other species (e.g., wedge-tailed sabrewing, an Emerald, in the mountains of eastern Mexico). Colony-like nesting in females is not common but is known to occur in a Topaz, and a couple of Coquettes, including the Andean hillstar, in which groups of females nest together in protected ravines in the Andes. Also, females in a few species (e.g., a Hermit and a Mountain Gem) nest close to each other in the mating territory of a male, who is presumed to be their mate. Regardless of the dispersion patterns of the sexes, hummingbird mating systems are polygynous with males mating with several females each season. Whether the system is also promiscuous with females mating with more than one male per reproductive bout is currently unknown. Paternity analyses using microsatellite DNA, a technique that is commonly used to assess paternity in many animals and plants, are needed to determine whether the two-egg clutches of female hummingbirds have one or two fathers.

As is well known, females are solely responsible for nest building, incubation, and the feeding of nestlings, although a few instances of male parental care have been reported. Hummingbird nests are compact structures composed of a variety of soft materials (especially spider webs) and are placed in a variety of places that usually, but not always, provide protection from rain and predators. Although they are often placed on horizontal branches or in the crotches of trees, they can also be placed on rock surfaces or, in the case of many Hermits, on the undersides of large leaves. Females typically build them quickly and well before they lay their clutch of two eggs. If they nest more than once a season, they build a new nest for each brood, as I've seen with an Anna's hummer that nested twice in our backyard in two years. In both years, she placed the first nest on a low branch in a Texas ebony tree and the second nest higher up on an oleander branch. Both nests successfully fledged all four nestlings in both years—an unusual occurrence since fledging success of 50 percent or less is more common in hummingbirds.

About five weeks are required for completion of one nesting cycle. Females lay their two eggs on two days and begin to incubate them once the first egg is laid. Incubation lasts about two weeks in many species and up to three weeks in the Andean hillstar. At hatching, the honeybee-sized babies are naked and blind and have wedge-shaped bills; by about twelve days they are well feathered. Females assiduously feed them a mixture of tiny insects and nectar about every twenty minutes in the case of the Anna that I've watched and photographed. The young birds and their bills grow rapidly, and they usually fledge in just under three weeks. At that time, their bills are long and thin and are nearly the same length as their mom's bill. The female Anna's that I've watched continue to feed their fledglings for a few days before they disperse, but in the tropics females may continue to feed their young for several weeks postfledging.

The breeding seasons of hummingbirds are correlated with the availability of their flowers. Females of many species, especially in the tropics but also in Temperate Zone, produce two broods each year. Anna's hummer begins nesting in December but most temperate breeders nest in the spring and summer. In Costa Rica, most species breed during the dry season (January to April) when many of their food plants are flowering. A second flowering peak early in the wet season (June to September) allows wet forest species to molt and build up fat reserves prior to a sharp drop in flower availability at the end of the year.

A detailed study of the long-billed hermit in the wet lowland forest of northeastern Costa Rica highlights the major events in the life and reproductive cycle of this common tropical hummingbird. (Note: when this study was conducted, this species was known as the long-tailed hummer, but its common name is now the long-billed hummer, whose cognitive abilities at this study site were described above.) This comprehensive study, which was conducted by my friends Gary Stiles and Larry Wolf while I was working with rodents and bats at the same site (the La Selva research station), is notable because it integrates this bird's life history with seasonal changes in rainfall and its food supply throughout the year. The study's subject is a relatively large (ca. 6 g), dull-colored hermit with a long, slightly decurved bill (ca. 37 mm); males and females are monomorphic (figure 8B). It is probably the most common of the twenty-two species of hummers in and around La Selva.

The climate in this region of Costa Rica is mildly seasonal and very wet—annual rainfall is about four meters. Lowest rainfall (but never really dry) occurs sometime between late January and late April; heaviest rainfall (As we all can attest! I well remember Gary and Larry telling me about frequently sliding down a steep trail called "Hans's Horror" on the way to one of their lek sites during the wet season) occurs in June to July and November to December when rivers are swollen and trails at La Selva are deep in mud (or were at the time of this study). Plants generally respond to the mild seasonality in this area by having two flowering peaks—a major one in the mild dry season and another less strong one at the beginning of heavy rains in June or July through September. The twenty-one species of plants known to serve as nectar sources for long-tailed hermits generally conform to this seasonal pattern. These species occur in ten plant families and fall into two groups regarding their flower morphology: species with long, curved flowers (mainly megaherbs in three families of monocots) and species with long, slightly curved or straight flowers (in the other seven families). At this site, the curved flowers of nine species of *Heliconia*, particularly *H. pogonatha*, were favored by hermit hummingbirds, which were their exclusive pollinators. The other flowers were pollinated by both hermits and non-hermits.

As is typical of hermits, adult male long-tails at La Selva occurred in four main leks containing twelve to twenty-three males apiece and located at least a kilometer apart. Within a lek, each male had a territory containing several singing perches on which it spent most of the day singing, displaying, preening, or sitting quietly throughout the breeding season, which ran from January to August. Feeding by these males occurred early in the morning when nectar levels in their preferred flowers were high. Lek occupancy and singing behavior declined strongly in September through November. During the breeding season, females apparently rarely visited the leks and then only for mating. Observed copulations were very fast and lasted only three to five seconds. Annual turnover among lek males was high, and only about 50 percent of males occupied a lek for two years. Most males live only one or two years in this population.

Females built their nests away from leks by placing them on the underside of large leaves and raised their babies alone. Female nesting behavior was not included in this study, so little is known about this aspect of this bird's life history. What we need to know includes how many clutches each female produces per season and what is their fledging success. Probable answers to these questions include females likely produce two clutches per season and the fledging success of each clutch is low as a result of predation by snakes. In addition, paternity analyses have not yet been conducted to determine the mating success of different lek males and whether the two-egg clutches have a single father. Judging from other lek-mating species, however, we can make three predictions about the mating success of males. First, it will be strongly skewed; only one or two males in a lek will garner most matings. Second, males with territories in the center of leks will garner more matings than other males. And third, males that display most vigorously will garner more matings than other males. Given their relatively short life spans, males of this species must be under intense pressure to gain a territory within a lek and try to attract as many females as possible in their first breeding season. At most, they will have only two breeding seasons to father any offspring.

HUMMINGBIRDS IN A COMMUNITY CONTEXT: HOW IMPORTANT IS COMPETITION FOR FOOD?

The number of species of hummingbirds that co-occur in the same habitat, at least seasonally, varies from a single species in eastern North America to about twenty-eight species in various South American localities. In tropical America, hummingbird species richness (S) varies with rainfall. Both lowland and montane dry habitats support far fewer species (typically a maximum of eleven species) than lowland and montane moist or wet habitats (typically a maximum of twenty-two to twenty-eight species). Hummingbird communities in the West Indies resemble those in dry habitats in the mainland in their low S (two to four species). As discussed previously, hummingbirds are mobile animals and hence their communities contain both nonmigrant (= principals of Feinsinger and Colwell) and migrant (= secondaries) species. In several well-studied sites in Costa Rica, for example, the percentage of migrant species ranges from 12 to 60 percent. The presence of substantial numbers of migrant species at many sites emphasizes the dynamic nature of hummingbird communities. Their composition is not static but changes throughout the year. Finally, in most hummingbird communities, the abundance of different species varies widely with only three to six common species—usually the principals—and many uncommon species in the richest sites.

Before looking in some detail at the main factors that determine the number of coexisting hummingbirds in communities, let's return to the "big picture" and use our detailed phylogenetic history of this family (figure 3) to see how phylogeny (i.e., the nine clades) influences patterns of hummingbird coexistence. In doing this, we can ask, do communities contain a high diversity of species in different clades but a low diversity of species within the same clade? We might expect this if species within clades, because of their similar morphologies (table 4), are likely to be strong competitors for food resources whereas species in different clades, because of their different morphologies, are less likely to be competitors. Catherine Graham and colleagues addressed this question by examining the phylogenetic composition of 189 communities along environmental gradients in Ecuador, which has one of the richest hummingbird faunas in the world. Their results indicated that communities in wet lowland habitats tend to contain species from different clades more frequently than expected by chance whereas montane communities and dry lowland habitats contained species from the same clade(s) more often than expected by chance. Overall, local communities do not contain sets of species drawn at random from the entire hummingbird fauna of Ecuador. Phylogeny and all that it implies regarding the biological differences that exist among clades is important in determining the composition of these communities.

Given that phylogeny matters, what biotic factors are involved in determining the number of hummingbirds that coexist together in a habitat? To start off, we might guess that hummingbird S might be positively correlated with the species richness (S) of their food plants. And we would be right. There is a significant positive correlation between hummer S and plant S across many communities. Dry habitats contain many fewer hummingbird flower species (and hummingbirds) than moist or wet habitats. This is hardly surprising because this trend is a general pattern seen in many kinds of life on Earth. Dry habitats are usually much less species-rich than wet habitats. This trend is easy to understand because it reflects differences between habitats in plant productivity, the ultimate source of energy and nutrients for animals. Plant productivity is obviously much higher in moist or wet habitats than in dry habitats. Net yearly production of new plant material in tropical forests, for instance, is about 1,800 g/m^2 compared with about 70 g/m^2 in dry habitats such as the Sonoran Desert—a twenty-six-fold difference.

Seasonal changes in flower availability are a major factor behind the dynamic nature of hummingbird communities. Many studies have shown that species that move between habitats or that undergo altitudinal or latitudinal migrations do so in response to seasonal changes in their flower resources. For example, a high-elevation (2,900 m) temperate habitat in Mexico contains a total of eight species of herbaceous hummingbird-pollinated plants whose main flowering season is May to October; peak flowering occurs in July. Flowers of these plants feed eight species of hummingbirds: two resident species, three altitudinal migrants, and three winter visitors that are long-distance migrants. Peak hummingbird abundance occurs in July during the flowering peak, but the greatest number of co-occurring hummingbird species occurs in November with the arrival of the winter visitors. Thus, not only is overall plant species richness within habitats important for hummingbirds, its seasonal distribution also plays an important role in the lives of these birds.

If plant species richness helps to determine hummingbird species richness in local communities, what about the role of interspecific competition in determining this diversity? Since flower nectar is usually a scarce energy source compared with, for example, fruit and insects, even in highly productive tropical forests, we might expect to see evidence for the effects of competition for limited nectar sources in the structure of hummingbird communities. If competition is an important factor, for instance, we might see differences among coexisting hummingbird species in their morphology and food choice—differences that should reduce their dietary overlap. The important role that phylogeny plays in determining the species richness of moist or wet lowland communities in Ecuador is a reflection of the importance of competition in determining which species can coexist in these communities. The kinds of hummingbirds that live together in many communities often include a variety of foraging styles, including understory trapliners and territorialists and canopy territorialists and generalists. These differences help to reduce overall levels of competition and allow several to many species to coexist based on the diversity of their food resources.

One place to look for these differences might be on islands in the West Indies where plant species diversity is much lower than in similar mainland habitats. On Puerto Rico, for example, three species of hummingbirds—two large, long-billed species of mangoes (the Antillean and Green mangoes) and the small, short-billed Puerto Rican emerald—are widely distributed. In the dry forests of southwestern Puerto Rico, these birds visit a total of thirteen species of tubular flowers, and they segregate by habitat and flower size. The two mangoes occur in different habitats and visit only long-corolla flowers whereas the emerald mostly visits short-corolla flowers whenever it co-occurs with one or the other larger species. If the two large mangoes were to co-occur in the same habitat, they would probably compete for nectar produced by long-corolla flowers whereas neither species is likely to compete for nectar with the smaller emerald when they co-occur. These results suggest that both competition (especially between the mangoes) and coevolution (between bill sizes and flower sizes) have been involved in determining the structure of these species-poor communities. These processes are obviously also very important in structuring more diverse mainland communities.

In addition to highlighting the importance of interspecific competition in the composition of certain communities in Ecuador, results of the Graham and colleagues study point to the importance of environmental filtering as an important factor in determining the structure of high-elevation hummingbird communities. Environmental filtering often occurs when biological or physiological conditions change along environmental gradients that cause some species to be selected against along the gradient. As an example, we've already seen that hermit hummingbirds are basically restricted to lowland habitats because they lack the physiological capacity and wing morphology to survive successfully at elevations greater than about two thousand meters. In contrast, members of the Coquette and Brilliant clades have evolved the capacity of operate successfully at high elevations, and these two clades are very well represented in montane communities. As mentioned earlier, members of these two high-elevation clades differ in their habitat associations with Coquettes living in open habitats containing plants of low stature and Brilliants living in forests containing tall trees. In addition to the physiological challenges that occur along environmental gradients, these gradients can also include changes in the kinds of plant resources, competitors, and predators that species face along the gradient and that can affect their survival.

CONCLUSIONS

All of the evidence that I've presented in part 3 underscores how truly unique hummingbirds are among all forms of life on Earth. Virtually every aspect of their biology differs strongly from all other birds. Three of their features—their very small size, their ability to hover (albeit at great metabolic cost), and their very heavy reliance on a relatively unusual food source (plant nectar)—have had a profound effect on all aspects of their biology. Despite the challenges that these features create, hummingbirds are one of the most successful families of birds in terms of their species richness and morphological diversity. This success implies that they are endowed with a very labile genetic-developmental system that has enabled them to evolve an impressive array of morphological, physiological, and behavioral adaptations. And during this process, they have become important pollinators of many species of new-world plants and have thus played a major role in the diversification of life on Earth. Hummingbirds truly are tiny glittering packets of cells (about 5×10^{11}) and trillions of atoms, continuously striving to hold back Schrödinger's entropy. They clearly are tough little birds that are totally unaware of their place in the vastness of our universe. But they certainly have played an important role in the evolution of life on Earth.

4

HOW TO BUILD A NECTAR BAT

AN OVERVIEW OF NECTAR BAT DIVERSITY AND EVOLUTION

When I tell people that I study the behavior and ecology of nectar-feeding bats, most of them are surprised to hear this because, as far as they know, all bats eat insects, especially mosquitoes. And they're partially right, of course. Of the twenty or so currently recognized families of bats, only two families—the new-world leaf-nosed bats (family Phyllostomidae) and the old-world flying foxes (family Pteropodidae)—contain flower-visiting species, and these species are in the minority in both families. No family of bats is exclusively dedicated to nectar feeding as are hummingbirds and a few other families of birds. And as far as mosquitoes go, very few bats are dedicated mosquito eaters. Most insect bats go for larger prey such as moths and beetles rather than tiny mosquitoes.

Bats are classified in the order Chiroptera and are evolutionary products of the radiation of mammals in the early Cenozoic era. Their oldest fossils are known from the Early Eocene about fifty to fifty-two million years ago. Temperatures were relatively high then and tropical forests full of a burgeoning array of flowering plants and a diverse assortment of insects covered most continents. These fossils reveal fully formed bats, so the evolution of this group of flying mammals must have occurred earlier than this. But how bats evolved from arboreal, shrewlike mammals into superb, sophisticated nocturnal fliers—only the third group of vertebrates (pterosaurs and birds are the others) to do so—is not yet fully understood. The recent description of the oldest known bat, *Onychonycteris finneyi*, from an Early Eocene fossil bed in Wyoming, suggests that this bat was capable of flap-gliding flight but could not echolocate. The ability to fly apparently predated the evolution of echolocation in early bats. Furthermore, the presence of claws on all of its fingers and its hind limb proportions suggest that *Onychonycteris* was capable of climbing along or under the branches of trees. Once bats were capable of fully powered flight early in their evolution, they retained claws on only their first and second fingers (only in flying foxes) and acquired the ability to echolocate.

As the only flying mammals, bats have a much more expensive mode of locomotion than nonflying mammals as I discuss below. Knowing this, it should not be surprising to learn that the metabolic machinery involved in the production of energy-rich ATP molecules has undergone strong selection in the evolution of bats (as it also has in hummingbirds). Genetic analyses of selection on both mitochondrial and nuclear genes involved in energy production (the so-called oxidative phosphorylation or OXPHOS genes) indicate that a higher proportion of these genes have undergone significant positive selection toward higher rates of energy production in bats than in other kinds of mammals. Thus, an elevated metabolic rate as well as selection for the evolution of highly modified forelimbs, among many other morphological changes, were involved in the evolution of bats.

Modern phylogenetic studies based on a combination of genetic data, morphology, and fossils indicate that the order Chiroptera is most closely related to a surprising group of mammals called the Ferungulata, which contains four orders of which three are certainly familiar to most of us: Pholidota (pangolins), Carnivora (dogs, cats, and their relatives), Perissodactyla (horses, tapirs, and rhinos), and Cetartiodactyla (including camels and their relatives, cows and their relatives, and, unexpectedly, whales). They are not closely related to primates or rodents. All these orders evolved in the northern supercontinent of Laurasia, and bats likely first evolved in North America. Most of our modern bats, though, appear to have had a Eurasian origin, like swifts and hummingbirds. They now occur everywhere on Earth except for the treeless regions of high latitudes in the northern hemisphere and Antarctica.

In addition to being the only flying mammals, modern bats are notable because most species use echolocation to navigate and hunt for prey at night. Only one family of bats—the flying foxes (family Pteropodidae)—lacks this ability (except for tongue-clicking species of *Rousettus*). Since they are related to certain families of bats that do echolocate, pteropodids must have lost this ability early in their evolution. I'll discuss possible reasons for this later in the book. Echolocation generally involves the production of high-frequency sounds (i.e., frequencies greater than 20 kHz and above our hearing range) for navigating, feeding, and communication. This sensory adaptation has been extremely important for bats for at least two reasons. First, it has allowed them in use dark caves and mines as safe roosts during the day. The ability to echolocate has allowed them to navigate easily in the total darkness of deep caves. And second, it has allowed them to become superb predators of night-flying insects. Large eyes and wide mouths have allowed the Strisores nightbirds (the ancestors of hummingbirds) to feed successfully on insects at night, but compared with bats, their hunting behavior is far less sophisticated. Some species of Strisores pursue flying insects from perches or pounce on them on the ground; many others "filter feed" on aerial "plankton" while on the wing. In contrast, echolocating bats eat a much more diverse array of food, including many kinds of insects, fish, other vertebrates, blood, fruit, and nectar, that they capture using an equally diverse array of foraging behaviors. Nonetheless, echolocation is critically involved in "capturing" virtually all of these "prey" items.

Because it is such an important sensory adaptation in the lives of bats, I think it is worthwhile to review briefly some of its more important features. First, although laryngeal (i.e., throat-produced) echolocation is absent in the megabats of the Pteropodidae, all other bats can echolocate. Within the remaining nineteen or so families, however, there resides a tremendous diversity of echolocation styles. Bats have not evolved only one way to echolocate; they have evolved multiple ways to do this. They do this with a combination of anatomical and neurological ways to produce, receive, and process high-frequency sounds. Universal mammalian features involved in this process include the larynx for sound production; external pinnae (ears) for sound reception; a tympanic membrane, three tiny bones in the middle ear, and a coiled cochlea for sound transmission; and auditory regions of the brain for sound interpretation. Ultrasounds produced by bats usually occur in the frequency range of about 20 kHz–100 kHz but sometimes as high as 200 kHz. As you might guess, a bat's typical frequency is related negatively to its size; tiny species produce sounds of much higher frequencies than larger species. The form of a bat's echolocation calls is related to where it forages. Species that usually feed in forest understories, for example, typically produce short pulses of relatively soft, low-intensity sounds that sweep down through frequency ranges of 10 kHz–100 kHz wide; these sounds often contain multiple harmonics and are therefore complex. This echolocation design is called frequency modulated or FM. It enhances short-distance prey detection in acoustically cluttered environments. In contrast, species that pursue prey in open habitats (e.g., above forest canopies) typically produce long pulses of relatively loud, high-intensity sounds of more or less constant frequencies. Not surprisingly, this echolocation design is called constant frequency or CF. This design enhances long-distance (i.e., distances of ≥ five meters) detection of insect prey. In both styles, the repetition rates of sound pulses increase dramatically as bats approach their prey. Not surprisingly, these are called feeding buzzes.

While most bats emit their echolocation sounds through their mouths (i.e., the so-called oral emitters), members of a few families, including phyllostomid bats, are nasal emitters. They usually fly with their mouths closed and issue sounds through their nostrils. Nasal emitters differ from oral emitters in producing low-intensity rather than high-intensity sounds. Nasal emitters are therefore sometimes known as "whispering bats" whereas oral emitters are known as "shouting bats." As an additional note, many recent photographs of phyllostomid bats show them flying, including approaching flowers, with their mouths wide open (figure 18). This suggests that they can be both oral and nasal echolocation emitters.

The family to which new-world nectar bats belong—Phyllostomidae—is a member of a "superfamily"—Noctilionoidea—containing seven families of related bats that occur in Madagascar, New Zealand, and the new-world tropics and subtropics. It has been suggested that this superfamily originated in South America, which was a member of the Gondwanan landmass, rather than in Laurasia, about fifty-two million years ago (figure 13). Phyllostomid bats are most closely related to two neotropical noctilionoid families of primarily insect eaters (i.e., Noctilionidae and Mormoopidae). They are derived members of this superfamily and likely separated from their closest relatives (family Mormoopidae, mustached bats) in the Late Eocene, about thirty-six million years ago. The family appears to have first evolved in North America, but, like hummingbirds, its greatest diversity now occurs in tropical South America. Also like hummingbirds, the family is mostly distributed in the lowlands (up to about 1,500 meters), but, unlike hummingbirds, it has undergone only a modest radiation in the Andes. I'll discuss possible reasons for this later.

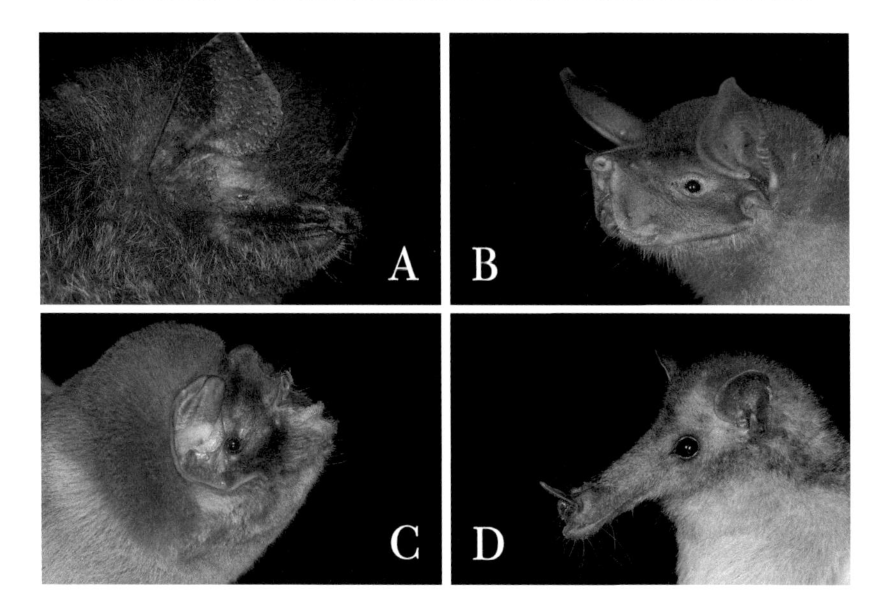

FIGURE 13 Four examples of Noctilionoidea bats: A. Thyropteridae (*Thyroptera discifera*), B. Noctilionidae (*Noctilio leporinus*), C. Mormoopidae (*Mormoops megalophylla*), and D. Phyllostomidae (*Musonycteris harrisoni*). *Thyroptera, Noctilio,* and *Mormoops* are from Central and South America; *Musonycteris* is from Mexico. Photo credits: Marco Tschapka.

Phyllostomid bats are called "leaf-nosed" bats because most species have a triangular flap of skin called a noseleaf above their nostrils. Reflecting the substantial diversity of food habits found in this family—insects, vertebrates, blood, nectar, and fruit—these noseleaves range from very small and reduced in size (in vampire bats) to large and spear shaped in certain insectivores. The noseleaves of frugivores and nectarivores are medium to small sized. All of these bats are nasal emitters, and it is thought that their noseleaves, which can be quite mobile, serve to reflect and provide directionality to sound pulses as they leave the nose.

Phyllostomid bats are notable for their diversity of food habits. In fact, this family is clearly the ecologically most diverse family of bats in the world. Its ancestral feeding mode is insectivory. But unlike most families of insectivorous bats, which capture their prey by aerial pursuit, ancestral and modern-day phyllostomid insectivores often capture their prey by gleaning them off vegetation and the ground. In doing this, they are exploiting a prey base that differs substantially from that of other insectivorous bats, and hence they are not in direct competition with them for food. From this ancestral feeding mode, eleven subfamilies of phyllostomids have evolved in the past thirty million years or so. These subfamilies are equivalent to the nine hummingbird clades but currently lack fancy or colorful names. They include five subfamilies that are basically insectivorous (plus carnivorous in one of them), one subfamily that eats blood (the vampires), three subfamilies that eat fruit, and two subfamilies that feed on nectar. The two nectar-feeding subfamilies are Glossophaginae (with 14 genera and 36 species) and Lonchophyllinae (with 5 genera and 18 species). They are not sister lineages, which indicates that nectarivory has evolved independently twice in this family. Overall, this family currently contains 216 species, which makes it the second most species-rich family of bats. Only the vesper bats (family Vespertilionidae) with over 300 species is larger.

A simplified phylogeny of the phyllostomids containing estimated dates of divergence of some of its major lineages is shown in figure 14. Its oldest subfamily (Macrotinae) dates from about 30 million years ago (Ma), and the two nectar-feeding subfamilies date from about 20 Ma (Glossophaginae) and 15 Ma (Lonchophyllinae). Glossophagine bats are widely distributed throughout Mexico, Central and South America, and the West Indies whereas lonchophyllines are primarily South American in distribution. It is interesting to note that unlike hummingbirds, whose oldest lineages (Topazes and Hermits) are primarily South American in distribution, the oldest phyllostomid subfamily (Macrotinae, which contains a single genus and two [and likely more] species), occurs in the southwestern United States, arid parts of Mexico as far south as Guatemala, and the West Indies. These bats weigh 12–19 g and have large ears and eyes, a long, membrane-bound tail, and soft, fluffy fur. I have studied these bats in the Sonoran Desert and the Bahamas and West Indies and have always been impressed by their gentle dispositions and their curiosity. During the day they live in caves and mines, and they have sometimes flown up to me and checked me out as I've entered their roosts. In the Bahamas and probably elsewhere in the Caribbean, these bats are notable for preying on large black witch moths (*Ascalapha odorata*, Erebidae) weighing well over 1 g and having a wingspan of over 20 cm. I've encountered piles of their large wings under *Macrotus* feeding roosts in Bahamian caves, where these moths are called "money bats." Many Bahamians believe that if one of them lands on you, you will soon come into money.

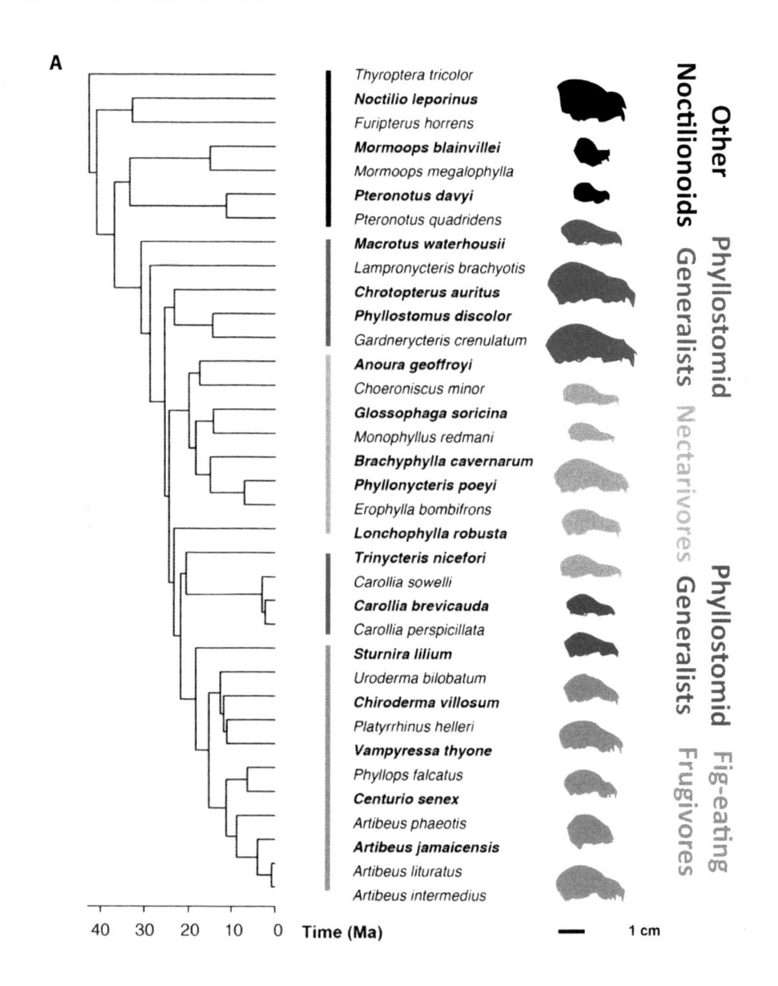

FIGURE 14 A dated evolutionary history of phyllostomid bats. There are two lineages of nectar feeders (Glossophaginae and Lonchophyllinae). From Mutumi et al. (2023) with permission.

COMPARISON OF NECTAR BATS AND THEIR ANCESTORS

The mustached bats (Mormoopidae) and phyllostomids are sister families, but, like swifts and hummingbirds, they are very different from each other (table 5). For one thing, Mormoopidae is a small family containing only two genera and about nineteen species. All of them are fast-flying insectivores with long, narrow wings. They are oral sound producers, and many produce long CF echolocation calls. They weigh 8–23 g, and their faces can be quite bizarre.

TABLE 5 A brief comparison of the major biological features of two sister families of bats: mormoopids (Mormoopidae) and phyllostomids (Phyllostomidae)

FEATURE	MORMOOPIDS	PHYLLOSTOMIDS
Taxonomic diversity	2 genera, 19 species	60 genera, 216 species
Geographic distribution	New-world tropics and subtropics and West Indies	New-world tropics and subtropics and West Indies
Size (mass) range	8–23 g	4–235 g
Morphology	Slender with long narrow wings and bizarre facial features; no noseleaves	Very diverse in size and shape; all possess a noseleaf of variable size and shape
Echolocation features	Oral sound emitters; species produce high-intensity ultrasonic calls; some calls are frequency modulated (FM) whereas others are basically constant frequency (CF)	Nasal sound emitters; most species produce low-intensity, frequency-modulated (FM) ultrasonic calls
Roosts and colony sizes	Most live in large, very warm ("hot") caves; colonies can contain tens of thousands of bats	Occupy many different kinds of shelters (hollow trees, logs, foliage, caves and mines, etc.); colony sizes are typically small (a few to a few thousand)
Diet and foraging	Strictly insectivorous; prey captured via aerial hawking	Diets are highly diverse (insects, blood, vertebrates, nectar, and fruit); insects captured via gleaning from vegetation and ground
Breeding systems	Typically harem polygyny	Variable: monogamy, harem polygyny, promiscuous
Reproduction	Litter size is one; most females breed once a year (monestrous)	Litter size is one; breeding frequency is variable; some species are monestrous, many others are polyestrous
Social behavior	Highly gregarious; often forage in mixed-species flocks	Most species are highly gregarious; a few species forage in groups
Migratory behavior	Nonmigratory?	Mostly nonmigratory; a few species are altitudinal or latitudinal migrants

The ghost-faced bat, for instance, has tiny eyes encircled by large cauliflower-shaped ears and a series of leaf-shaped fringes below its chin (figure 13). Other mormoopids also have tiny eyes and thick, flange-like lips with long mustache-like hairs projecting from their corners. Some of them also have wing membranes that meet along the dorsal midline, giving them a "naked-backed" appearance.

Most mormoopids live in large colonies that sometimes contain hundreds of thousands of individuals. They prefer to live in large, very warm ("hot") limestone caves. My friend Mandy Rodríguez, a bat biologist who lives in Puerto Rico where mormoopids are common, has described these hot caves as follows (pers. comm.):

> *Temperatures upwards of 35°C, relative humidity surpassing 90%, and levels of carbon dioxide that make these caves less than ideal for a leisurely stroll. It does not sound pretty, but it certainly is fascinating. Minuscule dipterans swarm your face, as they fly into the light beam of your headlamp. Boas, toads, tailless whip scorpions, and centipedes roam the cave, looking for an easy meal. You will struggle, as the muddy guano sucks your boots into this Hades, while a myriad of cockroaches and other invertebrates rush away from your path. You may even hear their chitinous exoskeletons brushing against each other. And what sounds like a distant breaking of ocean waves, is the flutter of thousands of bats, scared by your incoming approach. The movement of air, resulting from those thousands of wings, is the only relief in this hot environment, where sweat drenches your body, precluded from evaporating by the high humidity in the air.*

What a lovely home sweet home, right?

One rather bizarre aspect of these large mormoopid roosts in the West Indies is that their entrances can be encircled by a "necklace" of young boa constrictors that hang vertically from vegetation waiting to snag a bat as it leaves the cave among a swarm of roostmates each night. At a cave in Puerto Rico I once saw one of these snakes capture and eat two bats in less than an hour. As adults, these snakes cease hanging at cave entrances and become typical terrestrial vertebrate predators.

In contrast to mormoopids, phyllostomid bats, at least to my thinking, are much more handsome with far less bizarre faces. Many of the insectivorous species are basically medium-sized brown bats (usually weighing 20–30 g but up to 90 g) with relatively large ears and large noseleaves; large species have woolly fur. The spectral or false vampire bat is a true carnivore with a mouth full of large, robust teeth. Sometimes weighing over 200 g, it is the New World's largest bat and looks like an owl as it flies through tropical forests at night. It eats bats and other small mammals, birds, and reptiles. True vampire bats feed only on the blood of sleeping mammals and birds. They weigh 18–40 g and have short, sleek brown or gray fur, relatively small ears, large eyes, and nose pads rather than noseleaves. Their upper incisors are triangle shaped and have very sharp edges for cutting small divots in the skin of their victims. Their thumbs are robust and very long and serve as front feet for terrestrial locomotion. The nectar feeders are relatively small (up to about 30 g) and have small ears and small noseleaves and somewhat to strongly elongated snouts (figure 15). Finally, the fruit eaters, which range from 5 to 70 g in size, have short snouts, medium-sized noseleaves and large canine teeth for grabbing fruit. Most of them have bright facial stripes, and a few also have a dorsal stripe. They certainly are the most colorful of the over two hundred species of phyllostomids.

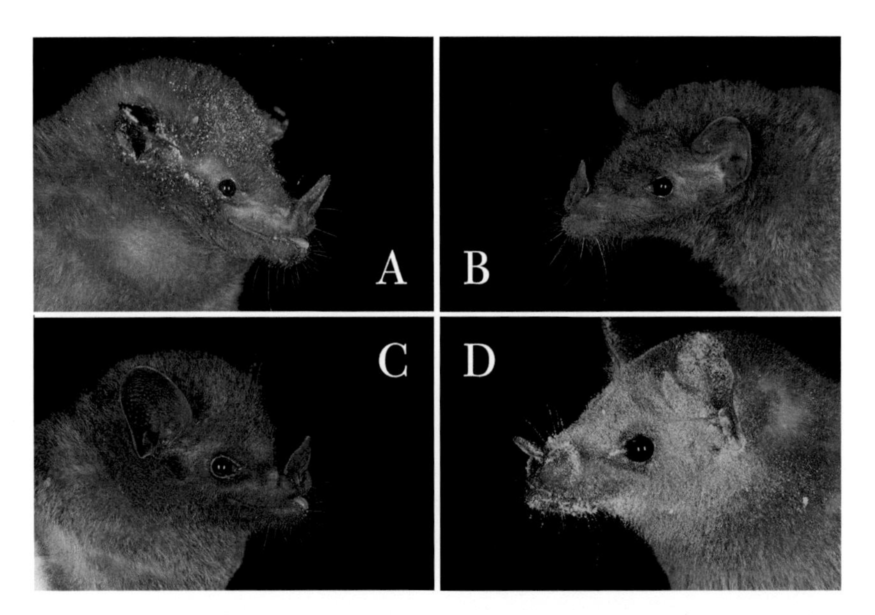

FIGURE 15 Four phyllostomid nectar bats: A. Thomas's nectar bat (*Hsunycteris thomasi*), B. Geoffroy's tailless bat (*Anoura geoffroyi*), C. common long-tongued bat (*Glossophaga soricina*), and D. lesser long-nosed bat (*Leptonycteris yerbabuenae*). Photo credits: Marco Tschapka.

Most phyllostomids live in caves or mines in colonies containing only a few hundred individuals. In the West Indies, they also live in hot caves with mormoopids. Others live in hollow trees or logs, and most fruit eaters roost in small colonies either in "tents" they construct by modifying large leaves in forest understories or in canopy foliage.

BASIC BIOLOGICAL AND ECOLOGICAL CHARACTERISTICS OF NECTAR BATS

Now that we've seen some of the basic characteristics of phyllostomid bats, let's begin to examine in some detail the characteristics of my favorite group, the nectar feeders. Because I've had much more experience with members of the glossophagine subfamily than the lonchophyllines, I will concentrate on them.

FIRST THINGS FIRST: THE MORPHOLOGICAL NUTS AND BOLTS OF BEING A BAT

Bats are perhaps the most distinctive group of mammals on Earth if only because they're the only mammals that can fly. Certain other groups of mammals—for example, colugos (order Dermoptera), marsupial sugar gliders, and so-called flying squirrels—have evolved the ability to glide from tree to tree using expanded lateral furred membranes between their front and hind feet, but they can't fly. Among mammals, only bats have evolved many of the morphological features involved in an aerial lifestyle. Two of these features include a small size and a very light skeleton (like that of birds). Most bats weigh far less than one hundred grams. Exceptions to this occur in the flying fox family Pteropodidae, in which a few species weigh over one kilogram. Glossophagine nectar bats weigh about 7–30 g; lonchophyllines weigh about 5–20 g. Hence all of them are small animals, like hummingbirds.

As you can guess, their small size has many important consequences for bats. In a recent paper Rubalcaba and colleagues used a modeling approach based on thermodynamic and aerodynamic principles to identify the optimal size and wing morphology of echolocating bats (the non-echolocators found in the megabat family Pteropodidae were not considered in this study). Their results indicate that size-related metabolic costs determine the minimum size of bats (as in hummingbirds) whereas the overall costs of thermoregulation during flight determine their maximum size. Echolocation imposes an additional metabolic cost and is also important in determining the size of most species of bats.

A quick glance at the skeleton or X-ray of a bat should convince you that these are very unusual mammals. What should strike you is the way the front half of their body with its large chest and large hands with four very long fingers that support the wing membranes dominates its body (figure 11). The back half with its relatively tiny legs, its backward pointing knees, and a thin tail that may or may not be present appears to be an afterthought in mammalian design, although tails can be an important component in prey capture in insectivorous bats. A compact skull usually containing a robust set of teeth and long slender arm and finger bones should provide a lasting visual impression of the essence of a typical bat.

Wings, which are large elastic membranes composed of two thin layers of skin derived from a bat's back and belly and stretched between the arms, highly elongated fingers, and legs and between the legs and tail (if present), represent the most unique anatomical features of bats. These membranes are obviously involved in flight and give bats a much more flexible surface for more maneuverable flight than the relatively rigid feather-dominated wings of birds. Many of the skeletal modifications found in bats, particularly those involved in the support and construction of the wings (e.g., a strong clavicle, very long fingers), are directly or indirectly involved with flight.

As in birds, the wings of various species of bats differ in their size and shape (i.e., their aspect ratio or [wing span^2/wing area]) and the amount of weight they support per wing area (i.e., their wing loading). High-aspect-ratio wings are long and narrow and are associated with fast, agile flight. Low-aspect-ratio wings are generally short and broad and are associated with relatively slow, maneuverable flight. Being fast but relatively maneuverable fliers, mormoopid bats generally have higher aspect ratios and more heavily loaded wings than most phyllostomids. According to an extensive analysis conducted by Ulla Norberg and Jeremey Rayner, the wings of glossophagine nectar feeders are quite distinctive within the Phyllostomidae. They have relatively low aspect ratio and heavily loaded wings, and large wingtips—traits that are associated with their exceptional hovering ability. Bats of the genus *Leptonycteris*, which are long-distance migrants, are an exception by having relatively high-aspect-ratio wings for fast, efficient open-country flight.

In addition to wing size and shape, the ears and tail membranes of bats are also involved in flight. Large ears and tail membranes can be sources of aerodynamic drag and hence increase the metabolic cost of flight. As might be expected given their taxonomic diversity, bats exhibit considerable diversity regarding the size and shape of their ears and the size of their tail membranes. In both mormoopid and phyllostomid bats, ears are generally small (with some spectacular exceptions in certain animal-eating phyllostomids) and tail membranes are often large. Glossophagine bats all have small ears, and many have broad tail membranes. Again, *Leptonycteris* bats are an exception by having a much-reduced tail membrane and no tail. A similar situation occurs in the Andean glossophagine genus, *Anoura* (which means "tailless"). Many fruit-eating phyllostomids also have reduced tail membranes and lack tails. In general, small tail membranes and the absence of a tail are associated with bats that commute directly from their day roosts to their feeding areas before beginning to feed. They are commonly called commuting bats. Among phyllostomids, many nectar feeders and fruit eaters are commuting bats, which accounts for their reduced tail membranes and the absence of tails in some species.

In addition to their wings, most recent studies of bat morphology have concentrated on their skulls and teeth. In the hummingbird section (e.g., pages 30–31) I discussed research examining the degree of integration of different aspects (modules) of their skulls during evolution. A similar approach has been taken regarding the evolution of bat skulls. Two major factors—allometry and modularity—are involved in the evolution of mammalian skulls. Allometry refers to the effect of body size on skull (and many other) features (see page 45). For example, large bats (like large mammals in general) have a longer rostrum (i.e., snout) than small bats. Modularity refers to the extent to which different skull features form correlated groups that are uncorrelated with other groups (e.g., the rostrum vs. the braincase). Of these two factors, it turns out that modularity has had a stronger effect on skull evolution in bats than allometry. Detailed analyses have revealed that the skulls of most bats contain five distinct morphological modules in their crania (i.e., the skull proper) and four in their mandibles (i.e., the lower jaw). As in birds, the existence of this modularity means that the many different features of bat skulls are not free to evolve independently from each other. Instead, skull features such as those associated with the anterior or posterior halves of the skull are constrained to evolve as single structural units. In phyllostomid bats, for example, morphological features associated with their rostrum (e.g., long in nectar feeders and short in fruit eaters) have evolved faster than those associated with their braincase.

In addition to the five cranial morphological modules that are basically involved in determining how strong these bats can bite (e.g., much stronger in carnivorous bats than in nectar feeders), the skulls of bats (and other mammals) also have three sensory modules—one each associated with olfaction, vision, and hearing—and it is of interest to know the extent to which these two sets of modules have evolved independently from each other. Have the morphological and sensory features of phyllostomid bats and their noctilionoid relatives evolved in highly correlated fashion or have they evolved independently of each other? A detailed analysis of these modules in noctilionoids has revealed that these two groups of modules have often evolved independently of each other and that features of the sensory modules, particularly an increase in the relative size of the olfactory and visual modules, have evolved faster in phyllostomid bats than in their insectivorous relatives. In fact, an increase in the size of these two sensory systems actually began before the evolution of early phyllostomids, which suggests that sensory evolution (i.e., the ability to find food via vision and olfaction rather than via echolocation) preceded morphological evolution in the skulls of these exceptionally diverse bats.

A plot of the location of three major feeding modes in echolocating bats—animals (including insects), nectar, and fruit—in 2D morphological space provides us with a general picture of evolutionary trends in their skull architecture (figure 16). This figure should remind you of how much hummingbirds differ morphologically from their Strisores ancestors (figure 7). Most bats fall within the animal-feeding group, which includes examples from eight families in figure 16. Insectivorous and carnivorous phyllostomids, including species of the ancestral phyllostomid genus *Macrotus*, occur in this large group. The skulls of phyllostomid nectar and fruit eaters clearly fall well outside of the animal-eating morphological space. Nectar bats have evolved very narrow skulls with a trend toward molars with reduced cusp areas. Frugivores have evolved very broad skulls as well as molars with much reduced cusps. It took about ten million years for these transformations to take place.

A comparison of the skulls of three phyllostomid bats—the basal genus *Macrotus*, an insectivore, and two genera of glossophagine nectar bats—further illustrates the changes that have occurred in the evolution of their skulls (figure 17). Like many insectivores, the *Macrotus* skull is relatively long

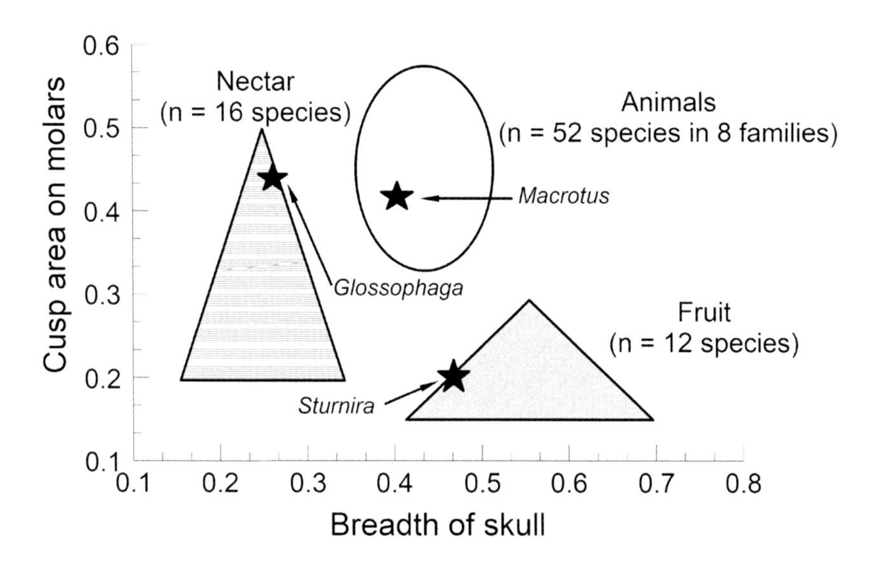

FIGURE 16 A 2D ordination of the skulls of three groups of bats based on their feeding habits. Basal phyllostomid genera in the groups are indicated by stars. Based on data in Freeman (1995).

and moderately wide. Its teeth are robust with sharp, pointed cusps on its W-shaped molars for crushing insect exoskeletons. In contrast, bats of the genus *Glossophaga* are probably close to the basal members of the nectar-feeding glossophagine radiation, and their skulls are delicate and somewhat more long-snouted with teeth that are markedly reduced in size and structure; their molars lack W-shaped (insect-crushing) cusps. The diet of these bats is quite generalized and includes some soft-bodied insects and fruit in addition to nectar. Bats of the genus *Choeronycteris* are among the most advanced glossophagines and have a very elongated rostrum and very small and simplified teeth. Their lower jaws lack incisors, a trait that they share with other very long-tongued nectar bats. Their diet is not well known but presumably is dominated by nectar. The skulls of other glossophagines also have somewhat elongated snouts and small, simplified teeth so the evolutionary trend in these bats is clear. You need a longer snout (and tongue) and smaller teeth than your ancestors if you're going to be a nectar bat. Their smaller teeth and long, slender jaws signify that in becoming a nectar feeder, these bats have mostly given up the ability to consume hard-bodied insects.

Although most of their teeth are reduced in size, phyllostomid nectar bats are notable for having relatively long and very sharp upper and lower canines.

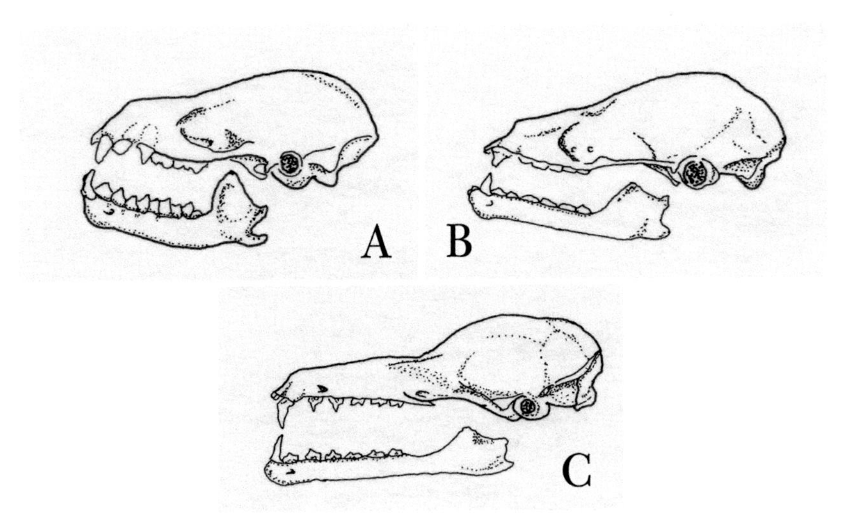

FIGURE 17 The skulls of three species of phyllostomid bats: A. a basal member of the family, the insectivore *Macrotus waterhousei* (skull length = 25 mm); B. a basal glossophagine, *Glossophaga soricina* (skull length = 20 mm); and C. an advanced glossophagine, *Choeronycteris mexicana* (skull length = 31 mm). Redrawn from Hall and Kelson (1959).

I can personally attest to this because many of the most painful bites that I have incurred during my tropical fieldwork came from the 10 g common long-tongued bat. Rather than being used for food gathering, however, these teeth are thought to serve as a brace in which the upper and lower canines interlock when the jaw is closed, forming a rigid arch through which the long tongue is extended. The lower jaws of these bats thus serve as a tongue-supporting structure rather than simply as a tooth-bearing structure.

Interestingly, the longest snouts in these nectar bats are not randomly distributed among its species. Instead, they occur, and have evolved independently at least three times (twice in glossophagines and once in lonchophyllines), in species that live in arid habitats where they feed extensively on the large flowers of columnar cacti. Nectar bats that live in moist or wet tropical habitats feed at smaller flowers and have shorter snouts than their arid-dwelling relatives (figure 18). Snout or rostral length in these bats is thus correlated with the length of the flowers they visit, just as the bills of hummingbirds reflect the corolla length of the flowers they visit.

Correlated with their longer snouts, nectar bats also have much longer tongues than their ancestors. The operational tongue length (i.e., how far

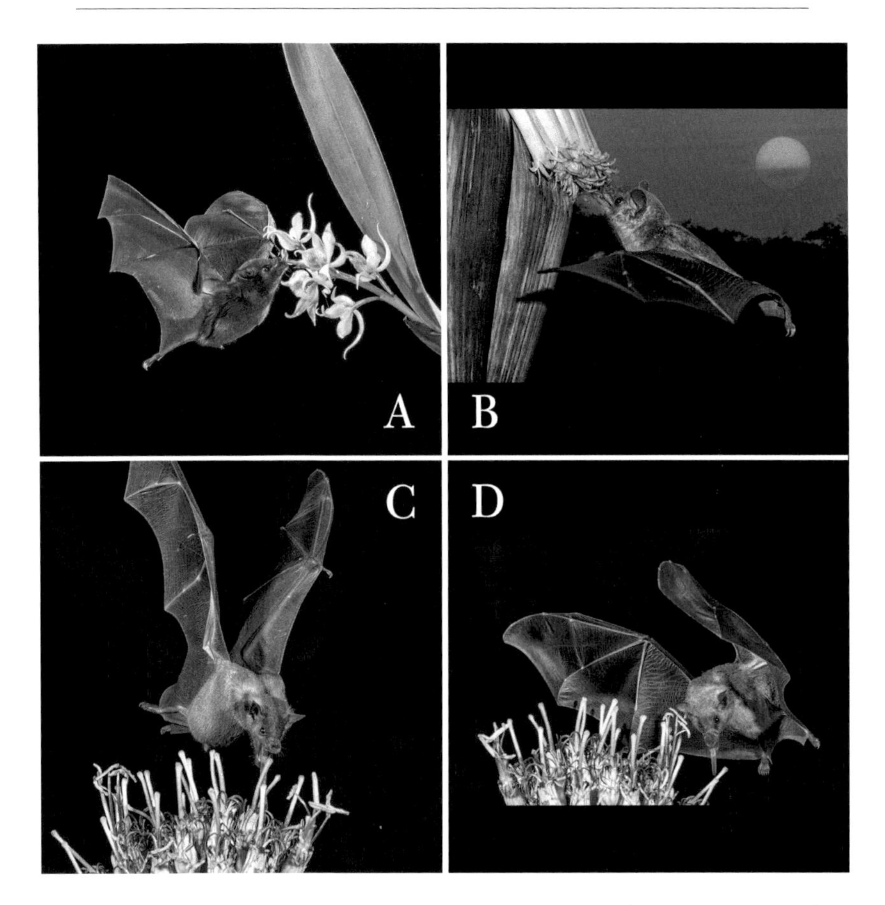

FIGURE 18 Four nectar bats in action: A. dark long-tongued bat (*Lichonycteris obscura*), B. common long-tongued bat (*Glossophaga soricina*), C. lesser long-nosed bat (*Leptonycteris yerbabuenae*), and D. Mexican long-tongued bat (*Choeronycteris mexicana*). A and B are from Costa Rica and Belize, respectively. C and D are from southern Arizona. Photo credits: Ted Fleming.

captive bats can extend their head and tongue into a narrow tube to reach sugar water) of nine species of glossophagines ranges from about 55 mm in a species of *Glossophaga* to 77 mm in *Choeronycteris* compared with only 24 mm for a fruit-eating phyllostomid. Even more impressive is the tongue of an Andean species of *Anoura*—the tube-lipped nectar bat—whose operational tongue length is a whopping 86 mm. This bat has the longest tongue relative to its size of any mammal. Incidentally, it lives in the same montane habitats as the sword-billed hummingbird, which has the longest bill (and tongue) relative to its size of any bird. Is this a coincidence? Please stay tuned! The tube-lipped nectar bat's tongue is so long that it has to be stored in a special cartilaginous tube in its chest cavity when it's not extended.

In addition to their length, the fleshy tongues of glossophagines are notable because they are tipped with a dense patch of tiny finger-like papillae that serve as a "mop" for quickly acquiring a dollop (about 10 mL) of nectar per flower (or hummingbird feeder) visit via rapid lapping of the tongue (figure 11). Unlike the micropump tongue extension and retraction system found in hummingbirds, however, tongue extension in nectar bats involves a vascular-hydraulic system in which blood is pumped into the fleshy tongue via an enlarged lingual artery and vein to extend it. Their tongues literally become erect while being extended.

Lonchophylline nectar bats, which evolved independently from glossophagines, have a somewhat different tongue structure and method of nectar uptake. They also have long tongues but lack the mop-like patch of papillae at their tips. Instead, their tongues have papillae-lined grooves along their sides. When extended into nectar, the tongue stays immersed rather than laps, and nectar flows through the grooves into the mouth via a combination of active peristalsis and capillary action. It has been suggested that this uptake method has evolved in lonchophyllines to acquire nectar from large flowers in which nectar is concentrated in one spot whereas the glossophagine "mop" has evolved to acquire nectar from smaller flowers in which nectar is more diffusely distributed. Despite these differences, nectar uptake efficiency appears to be similar in both kinds of bats.

HOW NECTAR BATS HOVER

Like their hummingbird analogues, phyllostomid nectar bats differ from most other bats in their ability to hover when visiting flowers. As we've seen, the oar-like wings of hummingbirds during hovering involve horizontal flapping in a figure eight motion. And their wings generate lift during both downstrokes and upstrokes. This results in energetically more efficient flight than nectar bats. In bats, in contrast, during level flight the downstroke provides most of the lift and the upstroke is basically a recovery, nonlift stroke. In addition, because their wings are built and operate differently from those of birds, hovering in bats involves somewhat different wing movements. As nectar bats hover, their wings undergo an upward turn during the upstroke, and the wing tip, which is enlarged in these bats, reverses itself to form a downward-facing triangle that provides lift as it undergoes a bent figure eight (propeller-like) motion that pushes air backward. The wing tip then returns to its normal position during the next downstroke. In addition, bats also have relatively larger wings than hummingbirds, which helps to compensate for their inability to produce significant lift during wing upstrokes. Finally, to partially reduce the forces of drag during hovering, nectar bats sometimes partially close their wings, something that rigid-winged hummingbirds cannot do (figure 11).

METABOLIC CONSEQUENCES OF SIZE, DIET, AND HOVERING

As we know, most bats, and this is certainly true of nectar bats, are small endotherms. Except during hibernation and daily torpor (see below), they operate at high daily temperatures, usually around 37°C, somewhat below the body temperatures of hummingbirds (about 41°C). And as we also know, being an endotherm has important metabolic consequences for them. For example, on a per-gram-of-mass basis, metabolic rates are highest in small endotherms (both birds and mammals) and decrease with body size, which means that both hummingbirds and nectar bats must feed at high rates to meet their daily energetic requirements. On the plus side, however, because they are small, their absolute daily energy requirements (a few milliliters of nectar) are not particularly high.

We already know that the metabolic rates and daily costs of living in nectar bats are high. How do these rates and costs compare with those of hummingbirds? Do they have similar daily costs of living? Little work has been done in this area for nectar bats, but available data suggest that, like hummingbirds, the daily costs of living as revealed by the doubly labeled water technique (see page 44) is likely to be high compared with nonhovering relatives. For example, a study of the 8.7 g Commissaris's long-tongued bat (another species of *Glossophaga*) in Costa Rica using this technique found that its cost of living is about 46 kJ per day compared with about 50 kJ per day for the 18.2 g fruit-eating silky short-tailed bat (a species of *Carollia*) in the same habitat. In other words, on a per-gram-of-body-mass basis, the daily cost of living for the nectar bat was nearly twice that of the larger fruit bat. On a mass-specific basis the daily cost of living in hummingbirds is about 2.5 times higher than similar-sized nonpasserine birds whereas it is about 1.6 times higher for nectar bats compared with nonnectarivorous bats. Thus, the daily cost of living of a

9 g hummingbird is about twice as high as that of a 9 g nectar bat. So it appears that nectar bats likely expend somewhat less energy per day than hummingbirds, but their daily energy expenditure is still high. Reasons for this include the relatively low energy rewards they obtain per flower visit (more on this below) and hence the large number of flower visits (i.e., hundreds) they need to make each day and the high cost of hovering during flower visits.

Both hummingbirds and nectar bats hover while feeding at flowers, so the question becomes, is the metabolic cost of hovering similar in both groups? Wingbeat frequency of hummers during hovering (30–60 per minute) is much higher than that of hovering nectar bats (about 9 per minute). As a result, the metabolic cost of hovering in nectar bats is only about one-half that of a similar-sized hummingbird. Nonetheless, when it is expressed on a mass-specific basis, the cost of hovering in nectar bats is among the highest values ever recorded for any mammal. Thus, being a hovering, nectar-feeding hummingbird or bat doesn't come cheap. Both groups have much higher daily living costs than their nonnectarivorous relatives.

As we've seen, hummingbirds have evolved an impressive suite of metabolic, enzymatic, and anatomical adaptations that allow them to rapidly convert the sugary nectar they ingest into fuel for their flight muscles. Have nectar bats evolved a similar suite of adaptations to fuel their high energy requirements during flight? The answer to this question is a resounding Yes. As a beautiful example of convergent evolution in unrelated organisms, many of the energy-related features found in hummingbirds are also found in nectar bats (and not in their insectivorous ancestors). For example, like hummingbirds, nectar bats have the ability to assimilate and digest sucrose and other carbohydrates in their intestines rapidly owing to high activity levels of particular enzymes (e.g., sucrase). Analysis of the suite of enzymes involved in this process indicates that there has been parallel evolution of sugar metabolism genes in both groups. Such evolution has not occurred in swifts and mormoopid bats, the sister groups of hummingbirds and phyllostomid bats, respectively. As a result, the assimilation efficiency of sugary foods eaten by both kinds of nectarivores is close to 100 percent. Furthermore, to support their high energy demands during forward flight and hovering, both groups have evolved large hearts, high oxygen transport capacities in their lungs, large flight muscles containing high densities of capillaries for oxygen delivery, and high densities of mitochondria in these muscles for producing energy-rich molecules (e.g., adenosine triphosphate, ATP). As a result, both kinds of nectarivores are able to quickly refuel their flight muscles with the sugar they ingest from nectar in a manner similar to the way fighter jet planes immediately use the fuel they receive from fuel tanker planes.

A diet rich in carbohydrates results in high blood glucose levels (hyperglycemia) while nectarivores are feeding. As we've seen, in hummingbirds, these high levels are matched by equally high rates of glucose metabolism during foraging. As a result, these birds do not display the usual negative signs of hyperglycemia. Although the biochemical details are not yet fully known, a similar situation likely occurs in nectar bats. They counteract their high glucose uptake with equally high levels of glucose metabolism. Experimental studies with the tailed tailless (!) bat indicate that insulin production, the usual way in which mammals regulate their blood sugar levels, might play a role in regulating blood glucose levels in this species but that glucose metabolism during flight likely is the major factor behind this.

Torpor is a common and important energy-saving metabolic strategy in hummingbirds, especially in species living under challenging climatic conditions such as those found at high elevations. Daily bouts of torpor in which both body temperatures and metabolic rates are reduced in hummingbirds typically occur at night and last for a few hours. These bouts can reduce a bird's total daily energetic needs by about 90 percent. How common is torpor in nectar bats? In their extensive review of the occurrence of torpor and hibernation in birds and mammals, Ruf and Geiser list seventeen species of hummingbirds but only one glossophagine bat, the 10 g common long-tongued bat, which is widely distributed in the neotropical lowlands, as undergoing daily bouts of torpor. Two additional phyllostomids, both fruit eaters, are known to undergo daily torpor, probably in response to low food levels. In contrast, torpor is much more common in both temperate and tropical insectivorous bats.

During torpor the common long-tongued bat's body temperature decreases to 21°C compared with about 37°C when it's active; its metabolic rate during torpor is up to 95 percent below its basal (or minimum) rate of metabolism, a considerable energy savings. Torpor is probably not a daily event in this bat, however, as it is in many hummingbirds. Its use of torpor to save energy depends on air temperatures and food availability. Its use of torpor is therefore facultative rather than compulsory. Rather than resorting to torpor to save energy, other glossophagine bats either switch to eating fruit when nectar levels are low or move to different habitats as many hummingbirds do. In fact, the common long-tongued bat is well known for switching to eating fruit and insects when nectar levels decline seasonally in its habitats. Regarding the use of torpor to save energy, therefore, the metabolic strategies of hummingbirds and nectar bats are very different.

One wonders why these two groups of nectar eaters are so different regarding their use of torpor to save energy. Why is torpor so uncommon in phyllostomid bats compared with many insectivorous bats and hummingbirds? An analysis of the metabolism of four species of mormoopid bats from Venezuela indicates that these "hot cave" bats also do not have the ability to undergo torpor. They maintain high body temperatures of 37°C–38°C over a wide range of ambient temperatures. Thus, it appears that both mormoopid bats and their sister family Phyllostomidae (with a few exceptions) are basically daily and year-round homeotherms that do not include the use of torpor in their metabolic physiology. If this is true, it would be particularly puzzling to find that the most abundant group of high-elevation nectar-feeding phyllostomids—bats of the genus *Anoura*—do not undergo torpor during cold days. This is known to be true in the tailed tailless bat and other species of *Anoura* (N. Muchhala, pers. comm.). The physiology of these bats thus differs strikingly from that of the hummingbirds living in the same montane habitats that routinely use torpor at night to save energy. An inability to undergo torpor is probably a reason why tailless nectar bats have not radiated in the Andes as extensively as have hummingbirds.

WATER, WATER EVERYWHERE REDUX

The diets of both hummingbirds and nectar bats feature a carbohydrate- and liquid-rich but protein-poor food source. We've seen that hummingbirds deal with this physiologically without highly specialized kidneys and by eating insects regularly to obtain protein and other nutrients. How do nectar bats deal with this? While some nectar bats do eat insects for their protein, others obtain it from a different source: the amino acids contained within pollen grains. Being mammals, nectar bats have fleshy tongues. Thus, it is easy for them to remove pollen grains from their fur with their long, flexible tongues while resting between foraging bouts. Once they've been ingested, these pollen grains are "digested" in the stomach, releasing their chemical contents, which include amino acids. When they are excreted, most of the pollen grains, whose exine (outer shell) is indigestible, are empty. For comparison, when pollen grains are experimentally fed to frugivorous phyllostomids, their contents are still intact when they are excreted, indicating that these bats lack the ability to "digest" pollen grains in their stomachs. Hummingbirds, in contrast, have cartilaginous nonfleshy tongues and cannot groom pollen grains off their feathers. And even if they do ingest some pollen grains, they cannot digest them to acquire their amino acids.

Like hummingbirds, nectar bats consume a large amount of watery nectar each night. For example, the common long-tongued bat consumes up to 150 percent of its body weight in nectar per night. As a result, the water economy of glossophagine nectar bats differs from most other mammals. In most mammals, their kidneys have evolved to conserve water and excrete excess electrolytes (i.e., salts) gained from their diets. The opposite is true for the well-studied desert-dwelling lesser long-nosed bat. Its kidneys need to excrete as much water and conserve as many electrolytes as possible. To do this, the structure of its kidneys differs from that of desert-dwelling insectivorous bats, including mormoopids, by having a small internal renal papilla (the site of maximum water resorption) and a very thick outer medullary cortex (the site of electrolyte resorption). As a result, these bats excrete copious amounts of very dilute urine, the most dilute of any mammal. In contrast, desert-dwelling insectivorous bats excrete very concentrated urine and need to drink water to remain in positive water balance.

HOW DO NECTAR BATS PERCEIVE AND INTERACT WITH THEIR WORLD?

Many aspects of the sensory worlds of hummingbirds and nectar bats are clearly very different. For example, being diurnal, hummingbirds live in a world full of bright colors whereas, being nocturnal, nectar bats live in a world that is basically grayscale and devoid of bright colors. In addition, having a poor sense of smell, the olfactory world of hummingbirds lacks an olfactory dimension whereas, having a well-developed sense of smell, nectar bats live in a potentially rich olfactory world. And, of course, the auditory worlds of these two groups are very different. As a result of these differences, we should expect these two groups to have sensory systems that differ substantially in their emphasis on particular senses.

BAT BRAINS AND VISION

In addition to their ability to echolocate, most bats are equipped with all of the usual mammalian sensory adaptations that deal with the many features of their visual, auditory, olfactory, taste, and tactile world. Also, like most mammals, they are nocturnal and have visual and auditory senses that reflect this. Because all of their sensory inputs ultimately end up in the brain, let's begin to explore the sensory world of nectar bats by considering major features of their brains. In discussing their brains, it is important to remember that neural tissue is energetically expensive to build and maintain and that this is especially important for flying vertebrates in which weight considerations are paramount. Thus, the evolution of the brains of bats are constrained both by their small body size and by the fact that they fly. Despite these constraints, some of the sensory systems of nectar bats have been subject to especially strong selection pressures.

We've seen that hummingbirds have brains that are relatively larger than most other birds. This is also true in nectar bats because it has long been known that the relative size of the brains of plant-visiting phyllostomids are larger than those of their non-plant-visiting relatives and other insectivorous bats. Frugivorous and nectarivorous phyllostomids must use their visual, auditory, olfactory, and spatial senses to locate, assess the quality of, handle, and consume fruits or nectar. As a result, we would expect to find that regions of their brains that deal with this sensory information to be relatively larger than similar regions in the brains of insectivores. These regions include the superior colliculi (visual), inferior colliculi (auditory), olfactory bulbs, and hippocampus (spatial memory). Detailed studies of these and other brain structures in phyllostomids support this hypothesis. In addition, an analysis of the morphology of three major sensory systems—olfactory bulbs, eyes, and the cochlea—has found that the size of the olfactory bulbs and eyes of nectarivorous and frugivorous species are relatively larger than insectivorous phyllostomids whereas insectivores have a relatively larger cochlea than nectarivores and frugivores. These results support the idea that the sensory worlds of the plant visitors revolve around olfaction and vision whereas those of insectivores revolve around echolocation. Interestingly, the ancestor of phyllostomids also likely had an enlarged brain and large olfactory bulbs and eyes, suggesting that it was likely to be omnivorous rather than strictly insectivorous. This suggests that phyllostomid bats were preadapted toward feeding on a variety of different kinds of food rather than just insects.

Jeneni Thiagavel and colleagues have recently reviewed the sensory and cognitive ecology of phyllostomid bats, and my account, starting with the visual systems of these bats, is based primarily on their review. Compared with nearly all echolocating insectivorous bats, including mormoopids, all phyllostomid bats have relatively large eyes, which tells us that vision—a longer-distance sense than echolocation—is especially important for them (figure 15). For example, vision is particularly important for long-distance orientation in the omnivorous greater spear-nosed bat. Experiments have shown that when confronted with conflicting sensory cues, some species give greater priority to visual rather than auditory cues. Experiments have also shown that some species can distinguish between different colors, sizes, and shapes using visual cues alone.

What is the visual acuity of phyllostomid bats? We've seen that because of their relatively small eyes, hummingbirds are slightly nearsighted and are not particularly sharp eyed. In contrast, many phyllostomid bats have relatively large eyes, and we expect them to have relatively good visual acuity. For instance, the California leaf-nosed bat (of the basal phyllostomid genus *Macrotus*), is a large-eyed, desert-dwelling insectivore that hunts its prey using visual cues. Its visual acuity is substantially better than two other desert insectivores—the big brown bat and pallid bat—of the family Vespertilionidae. Other large-eyed nectar- and fruit-eating phyllostomids are also considered to have better nocturnal vision than small-eyed insectivorous bats.

Because they are nocturnal, the retinas of phyllostomids are dominated by light-sensitive rods rather than color-sensitive cones. Nonetheless, the retinas of many phyllostomids do possess cones, and ultraviolet (UV) sensitivity has been demonstrated in many species, including insect- and fruit-eating species in addition to nectarivores. The ability to see UV reflectance in nectar bats could be particularly important for finding their flowers at night because it is known that many bat-pollinated flowers reflect UV light.

Color vision in mammals is determined by two opsin genes that code for long-wavelength (LW) (i.e., red-green) and short-wavelength (SW) (i.e., blue-UV) sensitivity. A recent survey of these genes in a series of phyllostomids has revealed that most species tested have retained a functional SW gene with the exception of vampire bats and the cave-dwelling lonchophylline Goldman's nectar bat. Loss of function of this gene has also been found in a variety of other cave-dwelling, nonphyllostomid bats, which suggests that living in dark caves has resulted in the loss of SW functionality. In contrast, a functional LW opsin gene has been found in all of the species tested, including insectivores, omnivores, nectarivores, and frugivores. Therefore, these results suggest that under certain light conditions, most phyllostomid bats possess the ability to see LW colors, as is the case in many other kinds of bats. These bats are not totally color-blind as their nocturnal lifestyle might suggest.

HOW IMPORTANT IS SCENT TO THESE BATS?

As we've seen, the olfactory bulbs of plant-visiting phyllostomid bats are relatively large, which suggests that the sense of smell is important to them. They also have a well-developed vomeronasal organ, which is lacking in birds, in the roof of their mouth that allows them to taste incoming air with their tongues. Many ripe fruits (e.g., figs) and many bat-pollinated flowers produce scents that we can detect, supporting the idea that bats are likely to use olfaction for finding their food. Discrimination experiments have shown that fruit-eating phyllostomids have a very acute sense of smell and the same might also be true of nectar feeders. Anatomical studies, however, suggest that because they have broader skulls and a larger nasal cavity, fruit eaters are more likely to rely on olfaction to find food than nectarivores, which have narrower skulls and smaller nasal cavities. Nonetheless, results of an experiment with two species of *Anoura* nectar bats in which they were required to discriminate between artificial flowers that either had or did not have a scent indicated that when the flowers were placed in a clutter of branches and leaves, as often occurs in nature, bats chose the scented flowers more often than scentless flowers. In the absence of plant clutter, bats chose both kinds of flowers with equal frequency. So at least in some circumstances, flower scent is an important cue that nectar bats use to find food.

In addition to finding food, scent is also important during reproduction in certain nectar bats. For example, in both the buffy flower bat, which roosts in caves in the West Indies, and lesser long-nosed bats in Mexico and Venezuela, adult males produce scents that they use in conjunction with other sensory cues to attract sexually receptive females. Males of buffy flower bats produce garlic-like secretions from glands above their eyes whereas *Leptonycteris* males smear a smelly mixture of bodily fluids on bare patches on their backs during the mating season to attract females. This olfactory strategy apparently works because females are known to prefer to mate with scented rather than unscented males. The term *eau de cologne* apparently is something quite different in bats than it is in humans.

As we've already seen, orally produced echolocation sounds are important in the lives of phyllostomid bats, particularly in insectivorous species. As they forage, nectar bats typically produce very short (1–2 msec), multiharmonic

FM calls of low intensity. In some situations such as deserts or other open habitats, however, they are known to produce louder calls that can be detected from up to 10 m away. In addition to aiding in orientation and avoiding objects while foraging, echolocation is used by nectar bats for flower discrimination, including their shapes, and approaches to flowers. Below I will describe how the lesser long-nosed bat uses echolocation in approaching large cactus flowers.

SPATIAL MEMORY

Having large brains implies that animals such as birds and mammals have advanced cognitive abilities. We've already seen that this is true in hummingbirds, which can assess the amount and renewal rate of nectar in their flowers, among other things. With this knowledge, they can adjust their foraging routes and timing of visitation to particular flowers, which certainly increases their foraging efficiency. Since nectar bats are faced with the same kinds of food gathering challenges, we expect them to also have excellent cognitive abilities. And their relatively large brains suggest that this is true. One of these challenges involves spatial memory, whose neural basis is located in the brain's hippocampus. This sensory feature includes remembering the location of nectar-producing flowers over several nights. Evidence that at least some nectar bats can remember the locations of flowers comes from experimental studies that show that these bats continue to remember the locations of energetically rewarding sources long after they have either been removed or relocated. Given the importance of finding flowers and their nectar, which can sometimes be widely scattered in tropical forests, every night, it seems safe to assume that an excellent spatial memory has been strongly selected for in these bats.

PUTTING IT ALL TOGETHER

How nectar bats use different sensory modalities to find food comes from detailed studies of the behavior of the lesser long-nosed bat, a migratory species that feeds primarily on the large cream-colored flowers of columnar cacti when it is in the Sonoran Desert in the spring and early summer. The sensory modes these bats use for finding and feeding at these flowers include vision, olfaction, and echolocation. The distances over which these modes operate can be ranked as vision > olfaction > echolocation. These bats can undoubtedly see large columnar cacti and their white flowers from a considerable distance (i.e., a hundred meters or so), even on moonless nights. So vision must be important in initially telling these bats where to feed. Cactus flowers produce a musky scent that is also likely detectable from a distance, especially downwind. Echolocation, in contrast, is effective as an information source about flowers from a distance of only a couple of meters.

In the first of two studies, Tania Gonzalez-Terrazas and colleagues used captive lesser long-nosed bats to examine the importance of echolocation plus scent vs. only acoustic information in choosing which target to visit. A large scent-producing cactus flower was one target and an acoustically visible but scentless acrylic hemisphere was used as the other target; a reward of a 17 percent sugar solution was added to both targets. Beginning about one meter from both targets, bats began to produce their characteristic echolocation calls, but bats only fed at the cactus flowers, indicating that a combination of both acoustic information and scent was needed to elicit feeding behavior. Interestingly, bats responded to the acoustic properties of the bell-shaped flower (which yields a more focused echolocation signal) and the hemisphere (a more diffused signal) with different approach behaviors. Bats headed into the cactus flowers very precisely whereas their approaches to the hemisphere were more variable. This supports the idea that echolocation signals from flowers allow these (and undoubtedly other) nectar bats to orient themselves very precisely when visiting flowers, even in total darkness.

As an aside, I have been using a trail camera to record lesser long-nosed bats visiting a saucer-shaped hummingbird feeder in my yard for several years. It never ceases to amaze me how precisely these bats are able to orient themselves and then insert their tongue into a 5 mm hole for 0.4 seconds as they hover in midair over the feeder. These feeders have none of the visual and acoustic characteristics of cactus or agave flowers, yet bats of all ages (e.g., juveniles and adults) are able to get a tongueful of nectar at feeders without mishaps. How do they do this?

In the second study, these researchers again examined the relative importance of scent vs. echolocation in feeding choices of captive lesser long-nosed bats. This time, however, cactus flowers, their scents, and acrylic hemispheres were used separately as potential targets in three experimental treatments: scent only vs. acoustic only cues, acoustic only vs. acoustic+scent cues, and scent only vs. acoustic+scent cues. Results indicated that bats reacted much more strongly and positively to the combination of both the acoustic and scent cues than to the two cues separately. These results provide strong support for the hypothesis that nectar bats use both acoustic and olfactory information while visiting these flowers.

Returning to my backyard feeders, how important is scent to these bats for finding and harvesting their nectar? Unlike bat-attracting flowers, hummingbird feeders are unlikely to be advertising themselves with scent. Sugar water is not volatile, so if it does produce a detectable scent, its odor plume must be very small. Out of curiosity, I gave my bats a choice of two identical feeders arranged side by side for six nights—one contained sugar water and the other contained tap water—and filmed their behavior with a trail camera. Although bats approached both feeders many times, they fed at the "nectar" feeder 309 times compared with only 16 times at the "water" feeder. The bats were clearly able to discriminate between the two kinds of feeders using some kind of cue. I suspect the cue must be an olfactory one, but further work is needed to test this hypothesis.

HOW DO NECTAR BATS FEED?

The lifestyles of hummingbirds and nectar bats differ profoundly in a way that certainly affects their foraging behavior. Hummingbirds are solitary animals that spend their "downtime" (i.e., at night) roosting near their feeding areas. Consequently, they spend far less time and energy commuting from their sleeping roosts to their feeding areas than they spend foraging. In contrast, most phyllostomid bats are gregarious and spend their "downtime" roosting with conspecifics in caves or cave-like structures (e.g., mines, culverts, hollow trees, and buildings). This is certainly true of most nectar bats. A few of these species roost in small colonies in tree cavities. The cave dwellers usually live some distance (up to nearly 100 km!) from their feeding grounds and therefore spend a substantial amount of time and energy commuting to these areas before beginning to feed.

As a result of these roosting differences, the time and energy budgets of hummingbirds and nectar bats are strikingly different. For example, a study of the time and energy budget of the purple-throated carib on the Caribbean island of Dominica indicated that males of this year-round territorialist spent 63–85 percent of the time simply sitting and 4–18 percent of the time actively foraging; they spent very little time hawking insects or defending their territories. The energetic costs of sitting and foraging were about 27–58 percent and 26–64 percent, respectively, of daily energy costs. Similar values have been reported for the broad-billed hummingbird in Arizona. The occurrence of commuting behavior was not mentioned in either of these studies, so it is unlikely to be a significant component of their energy budgets.

In contrast, our radio-tracking study of the lesser long-nosed bat in Sonora, Mexico, reported that each day this bat spent about 110 min commuting (over 20 km one way), 190 min foraging, 60 min resting in night roosts, and 1,080 min in their day roosts. The estimated energetic values of these activities were 6.60, 11.13, 1.52, and 20.90 kJ, respectively, for a total of 40.2 kJ per day. These bats therefore spend about 8 percent and 13 percent of their time commuting and foraging, respectively, each day. In terms of their daily energy expenditure, they spend 16 percent and 28 percent commuting and foraging, respectively. These and other nectar bats therefore partition their daily time and energy budgets very differently from those of hummingbirds.

As we've seen, hummingbirds use two principal behavioral strategies to harvest floral nectar: traplining and territorial defense. Of these, traplining, which occurs most commonly in Hermits, appears to be less common than territoriality. Do nectar bats use these two strategies, and if so, under what circumstances is one strategy used rather than the other? Unlike hummingbirds, territoriality appears to be very uncommon in glossophagine nectar bats. It has only been reported in a couple of instances: once at flowers of a species of *Agave* in Colombia and once around inflorescences of an understory palm in Costa Rica. Instead, most species appear to be trapliners, in which they visit a set of flowers of the same or different species along a spatially defined route each night. When it was traplining rather than defending *Agave* flowers in coastal Colombia, for instance, the common long-tongued bat visited several species of plants that occurred in areas of 35–51 ha (about 88–128 acres) each night. Distances between feeding sites in these areas were typically 150–250 m, and sometimes up to 1,450 m, apart. Traplines of different individuals overlapped with each other so that up to four bats could be seen simultaneously visiting the flowers of one plant without displaying aggressive behaviors. In this and other species of nectar bats, a strong spatial memory undoubtedly helps them make a traplining foraging strategy successful.

In my account of the foraging behavior of hummingbirds, I asked whether they are optimal foragers that harvest energy either as time minimizers or as energy maximizers (see page 80). Can these same concepts be applied to nectar bats? From what we know about the foraging behavior of a few species of nectar bats, it is difficult to think that they are foraging in an optimal fashion as defined in terms of time minimizers or energy maximizers. This is likely because of the long distances these bats often travel from their day roosts to their feeding areas and the large distances they travel between flowers within these areas. For example, our calculations have indicated that lesser long-nosed bats feeding at the flowers of columnar cacti in the Sonoran Desert could meet their daily energy needs by restricting their visits to flowers on only one or two plants in an area of less than one hectare (about 2.5 acres). Instead, they visit at least one hundred flowers over an area of 100–250 hectares each night. Distances between feeding sites can be 100 m or more. And before beginning to visit these flowers, they spend a couple of hours continuously flying around their feeding areas, presumably searching for open flowers and waiting for them to accumulate enough nectar before they are profitable to visit. Instead of conserving as much energy as possible during their feeding periods, as many hummingbirds do, these bats appear to be profligate in their energy expenditure each night. One possible reason for this is that, because of their high intake of sugar each night, they need to undergo strenuous exercise to burn off excess energy to avoid the dangers associated with hyperglycemia (i.e., a diabetic condition).

Further evidence of the profligate nature of energy expenditure in the lesser long-nosed bat comes from my observations of these bats visiting hummingbird feeders in my yard in Tucson. Using a trail camera, I have recorded the visitation patterns of bats to feeders all night over two years. To my surprise, I quickly discovered that bats continued to visit them for about three hours after they had been drained each night. It was common to see them insert their long flexible tongues into an empty feeder, apparently searching for a few molecules of nectar water. Several volunteers in our community science project (see the paper by Fleming et al. [2021] in Notes and the Bibliography) have also reported observing the same thing at their feeders. I currently have no idea why they do this but know that this cannot qualify as "optimal behavior." In fact, because these bats continue to burn energy without any apparent energetic reward, it seems to be downright "unoptimal." Nonetheless, there must be some benefit to this behavior. Otherwise, it should be strongly selected against. We simply don't yet know enough about the foraging behavior of this bat to know what that benefit is.

As a final example of the profound difference between the foraging behavior of hummingbirds and nectar bats, during the day my backyard hummingbird feeders are vigorously defended in territorial fashion by one or more species of hummers. At night, the same feeders are visited by lesser long-nosed bats in groups containing up to ten or more individuals. Instead of fighting among themselves, however, these bats queue up to take turns visiting a feeder in rapid succession. When not actually feeding, they fly in and out of my yard in "follow the leader" fashion. In contrast, when hummingbirds aren't feeding, they simply sit on a perch near a feeder, singing and constantly looking around for intruders approaching their territories. As a result, they spend much less time on the wing and burning energy each day than lesser long-nosed bats.

HOW DO NECTAR BATS MAKE BABIES?

We've seen that hummingbird moms are totally responsible for building nests and raising one or more clutches of two babies each year. Males are simply sperm donors and provide no paternal care. Do phyllostomid nectar bats have the same reproductive characteristics? One important difference between hummingbirds and bats in general is that litter size in bats is almost always one compared with a clutch size of two in hummingbirds. As a result, the reproductive output of nearly all Temperate Zone bats is one baby per year (i.e., a monestrous condition). This, however, is not true of most tropical bats that are more likely to be polyestrous than monestrous, probably because they live in more salubrious climates where hibernation is unnecessary.

Within phyllostomid bats, both monestrous and polyestrous species are known; mormoopids typically are monestrous as are basal phyllostomids in the genus *Macrotus*. Monestrous phyllostomids include species that are large, carnivorous or insectivorous, migratory, and/or island dwelling. Seasonal bimodal polyestry in which females undergo two pregnancies a year is especially common in fruit-eating species. Both monestry and polyestry appear to be common in glossophagines; no reproductive data are apparently available for lonchophyllines. Basal glossophagines such as species of *Glossophaga*, which occur in the lowland tropics, are mostly polyestrous whereas species of *Anoura*, which are montane, are monestrous. Species of *Leptonycteris* and the Mexican long-tongued bat, which are migratory, are monestrous.

As in hummingbirds and most bats, paternal behavior is lacking in virtually all phyllostomids with one exception. This occurs in the large monogamous carnivore—the spectral bat—in which both parents forage separately, share food with each other and their offspring, and babysit while their mate is foraging. In harem-polygynous species (see below), males have virtually no contact with their offspring, even when they are being nursed within their harems. In other species, males and females live in different roosts during the maternity period, precluding any male parental behavior. The extent to which offspring continue to have contact with their moms once they are weaned is poorly known in these bats. In many mammals, young males disperse from their natal sites soon after weaning, which severs any maternal-offspring bond. Young females, however, sometimes remain in their mother's harem in harem-polygynous species or in their natal roosts. In general, though, postweaning female-offspring connections are not strong in these and other kinds of bats.

The reproductive biology of the migratory lesser long-nosed bat appears to be unique among phyllostomid bats. On the Mexican mainland, it has two reproductive demes (interbreeding populations) that are not genetically isolated from each other. In one deme, mating occurs in southwestern Mexico in November and December after which many females migrate about 1,000 km north to the Sonoran Desert of northwestern Mexico and southwestern Arizona to give birth in large maternity colonies beginning in mid-May. These births coincide with the flowering and fruiting seasons of three species of columnar cacti. In the other deme, mating occurs in central Mexico in May and June and births occur in southern Mexico in December and January, which are peak flowering months for the tropical dry forest trees whose flowers they pollinate. Bats living in Baja California appear to have a third birth period that occurs midway between those of the two mainland demes.

Whereas the reproductive cycles of hummingbirds are relatively rapid (i.e., fledging occurs in about three weeks after an incubation period of about two weeks), those of many phyllostomids are relatively slow. Gestation typically lasts about four months in polyestrous species, and babies are large; they typically weigh about 30 percent of their mom's weight at birth. (Imagine giving birth to a thirty-pound baby!) Females then nurse their babies for an additional month and a half for a reproductive cycle that lasts for about five and a half months compared with five weeks in many hummingbirds. These differences highlight a major difference between the life histories of hummingbirds and nectar bats. Hummers literally live life in the reproductive fast lane whereas bats in general live life in the slow lane. As a result, bats have much slower life cycles than nonflying mammals of a similar size.

WHAT'S THEIR GREGARIOUS LIFESTYLE LIKE? HOW SOCIAL ARE NECTAR BATS?

Except for lek-mating species—primarily Hermits—most hummingbirds lead solitary lives and cannot really be considered to be "social" in the sense of having regular, nonaggressive close physical contact with each other. They do vocalize with each other, which certainly is a form of social interaction. But except for a few Andean species, they don't huddle together to keep warm; they don't preen or groom each other; they don't lead each other to good feeding sites, and so forth. Females certainly care for their offspring before and, to a certain extent, after they fledge, but this is a short-lived association. Courtship and mating are obviously social behaviors, but again, although male courtship displays can sometimes be elaborate, these are short-lived events. No long-lasting pair bonds occur between males and females.

In contrast, as we've seen, most phyllostomid bats, including nectar feeders, live in colonies ranging from a few (e.g., in Mexican long-tongued bats) to many thousands of individuals (e.g., in lesser long-nosed bats). Within their day roosts, they usually huddle together in compact clusters that can contain dozens of individuals. These clusters certainly have a physiological benefit. Lab studies have shown that bats can save energy by huddling together. Bats are also constantly grooming themselves while clustered, but they seldom appear to groom each other. Allogrooming is much less common in these bats than autogrooming. While it is known that females of certain phyllostomid bats (e.g., the omnivorous greater spear-nosed bat) that live together in harems share their feeding areas and lead each other to them, it is not yet known whether any nectar bats do this. If any nectar bats do this, I suspect it will be found in the ubergregarious lesser long-nosed bat. They certainly forage in groups in the Sonoran Desert, but the dynamics of these groups (e.g., how long or short lived are they?) are currently unknown.

Colonies of phyllostomid bats, however, are usually not simply random aggregations of individuals. There is a definite social structure to these aggregations. This structure is typically based on their mating systems, which range from monogamous family groups through harem-based polygyny to apparently large-scale promiscuous systems. In their review of bat mating systems, Gary McCracken and Gerald Wilkinson recognized three basic configurations: (1) single male/multifemale mating groups (i.e., harems); (2) multimale/multifemale mating groups; and (3) single male/single female mating groups (monogamous families). They further distinguished subdivisions within these configurations as being stable year-round or as seasonal associations.

At the time of their review (in 2000), the mating systems of no phyllostomid nectar bats was known, and this situation is still basically true. Among fruit-eating phyllostomids, seasonal harem polygyny appears to be the common social structure and it seems reasonable to suggest that this may also be true in at least some nectar-feeding species, but this is not yet known. Harems do not appear to occur, however, in the ubergregarious lesser long-nosed bat in which adult males and females roost apart for most of the year. The two sexes come together for mating for only a couple of months each year. At these times, it appears that these bats form multimale/multifemale groups in which matings are likely to be promiscuous. In contrast, mating in the West Indian and Bahamian buffy flower bat involves a lek-like social system in which adult males attempt to attract females for mating using visual (wing flapping), olfactory (garlic-like scents), and auditory cues (both low- and high-frequency calls). During their annual mating season (late November to mid-January in the Bahamas), these bats form multimale/multifemale colonies in which mating appears to be promiscuous.

I've spent a fair amount of time in a cave in Costa Rica watching bat behavior, mostly in the frugivorous Seba's short-tailed bat. In this species, older adult males stake out a small area on the cave ceiling and defend that area, which contains a group of females (a harem), against other male intruders. They do this by shaking their wings at and boxing with these males. All of this is done silently (to us). But this experience did not prepare me for what my graduate student Kevin Murray and I witnessed when we began to study the social behavior of the buffy flower bat on the island of Exuma in the middle Bahamas. This rather omnivorous but otherwise nectar-feeding bat is a typical cave dweller, and its population numbers in different caves during the mating season on Exuma range from 50 to 350 individuals with a sex ratio of 1:1 adult males and females.

We began our behavioral observations in one dark cave, aided by a Sony digital camcorder equipped with supplementary infrared light, on December 18, 2003. My field notes from that day describe what we saw and heard as follows: "In the dark, three sounds dominate the scene: loud, occasional bat flights [not caused by bats flushing]; the softer, continuous whirr of wing-buzz displays; and loud 'tock' sounds (like tongue clicks)." What we were hearing and eventually seeing via our recordings was males that were calling audibly, fanning their wings constantly (to waft away the garlic scents produced by scent glands above their eyes), and occasionally making short, looping flights in front of their display area. At first glance, we thought we were seeing lek behavior for the first time in glossophagine bats. But further results based on a paternity analysis to determine who was fathering babies indicated that nondisplaying males were as likely (or more likely) to father offspring than displaying males. This does not happen in classic vertebrate leks in which only lek males father offspring. Nonetheless, finding a species in which some males are displaying so vigorously during (and after) the mating season was totally unexpected. Later on when we visited many Bahamian islands, we commonly saw these displaying males camped out in the tops of bell-shaped solution cavities both during and after the mating season.

HOW DOES THE COMMUNITY STRUCTURE OF NECTAR BATS COMPARE WITH THAT OF HUMMINGBIRDS?

As we've seen, hummingbird communities can contain up to 28 species that differ significantly in body size, bill size and shape, and foraging behavior. These differences plus interspecific territorial defense result in a significant partitioning of floral resources in species-rich hummingbird communities. Although there are far fewer species of phyllostomid nectar bats compared with hummingbirds (about 56 vs. 352 species), how does the structure of their communities compare with those of their diurnal avian counterparts in terms of species richness and abundance? As we will see, there are both similarities and substantial differences in communities of these two groups.

I need to add a few words about the word *community* here. In ecological parlance, *community* can be defined as a group of interacting populations based on the nature of their interactions. As it is often used, *community* and its structure refers to a relatively large group of animals such as a bird community or a mammal community that includes a variety of kinds of interactions. More inclusive groups such as nectar-eating birds or fruit-eating mammals are often called *guilds* or *assemblages*. I will continue to use the word *community* here because most of us can visualize its meaning more clearly than the more inclusive terms. But in actuality, the sets of hummingbirds or nectar bats that live together in a habitat that I'm discussing in this book are better described as assemblages rather than communities.

In 2005, my colleagues and I published an extensive review of the community structure of hummingbirds and nectar bats. Similarities in their species richness (S) include the following trends. In both groups, highest species richness occurs in northwestern South America and southern Central America, especially in the mountains for hummingbirds but not for bats. Species richness within communities decreases from the equator toward more northern or southern latitudes—a general pattern seen in much of life on Earth. Tropical habitats almost always contain more of your favorite kind of organisms than temperate habitats. Within habitats, S is positively correlated with the species richness of their food plants and annual rainfall in both groups. The diversity of their food species obviously plays an important role in determining the number of co-occurring species of nectarivores. Reflecting the much higher overall diversity of hummingbirds, their communities contain about four times as many species, on average, as nectar bats. Average number of hummingbird species per mainland habitat is 11.5 species (range = 3–28) compared with only 3.1 species (range = 1–6) in bats. Species richness of both groups is also much lower on islands in the West Indies than on the mainland (averages of 2.8 vs. 1.9 species in hummers and bats, respectively). Furthermore, although species richness in both groups decreases with elevation in mainland communities, it does so much more slowly in hummingbirds than in bats. Montane moist forests contain an average of about 16 hummingbird species compared with only 2.4 species of nectar bats.

Compared with other potential food sources, including insects and fruit, nectar is a scarce resource, and abundances and population densities within communities of both hummingbirds and nectar bats reflect this. For example, in phyllostomid bat communities (i.e., all of the phyllostomids that live in a habitat), species of frugivorous bats outnumber species of nectar bats by a factor of about 5.6; their abundances differ by a factor of a whopping 32.1. The reason behind these large differences is easy to visualize. Simply picture a fig tree containing thousands of ripe fruits (and swarming with fruit bats each night) compared with a vine containing a few flowers offering very small amounts of nectar per flower (and visited by only a couple of nectar bats each night). Similar trends occur in tropical bird communities: frugivores greatly outnumber nectarivores in number of species and abundance.

In addition to their much greater overall species diversity, it is not surprising to know that the morphological diversity that you can see in a typical hummingbird community far exceeds that of the morphological diversity of nectar bats in that community. A general feeling for this can be seen in figure 19 in which I've plotted data on a measure of overall size (on the x-axis) and bill or jaw length divided by wing or forearm length (i.e., a measure of relative jaw or beak size) for many *genera* (not species) of hummingbirds and nectar bats (on the y-axis). Figure 19 shows that the morphological space occupied by hummingbirds far exceeds that of nectar bats. This is true of communities as well as the overall faunas depicted in that figure. The two hummingbirds that

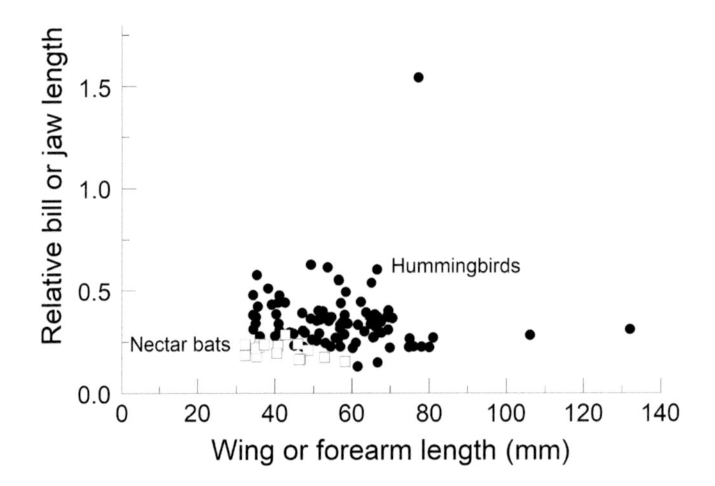

FIGURE 19 A 2D ordination of the size (x-axis) and bill or jaw length relative to forearm length (y-axis) in genera of phyllostomid nectar bats and hummingbirds. Based on data in Fleming et al. (2005).

greatly exceed other hummingbird values include the sword-billed hummer with its exceptionally long bill and the Andean giant hummer with its very large overall size. Compared with the exuberant range of sizes, plumages, and ornaments that occur in many hummingbird communities, nectar bats that live together are usually very similar in size and appearance and are downright dowdy compared with hummingbirds (figure 18).

Next, to illustrate what happens over a year's time in a community of nectar bats, let's go to a lowland tropical wet forest in Costa Rica, the same site where Larry Wolf and Gary Stiles studied long-billed hummingbird leks many years ago. Whereas the hummingbird community at La Selva contains about twenty-two species, only four species of nectar bats live there, and only two species are year-round residents. The other two species are altitudinal migrants that occur at La Selva only in October through February, a time of maximum flowering for nectar bats in this forest. The four species include three glossophagines (Commissaris's long-tongued bat [8.8 g], dark long-tongued bat [7.0 g], and Underwood's long-tongued bat [7.6 g]) and one lonchophylline (Goldman's nectar bat [15.8 g]). The two year-round residents (Commissaris's and Underwood's bats) differ in abundance and roost either in hollow trees or in the buttresses of fallen trees, respectively. They also differ in the relative length of their rostrums and wing shape. As a result, Underwood's bat appears to be a more specialized feeder and faster flier than Commissaris's bat, which switches to eating fruits when flowers are scarce. Plants providing nectar and pollen for these bats include nine species of trees, nine species of epiphytes, hemiepiphytes, and vines, and one understory palm. The flower diets of these bats are broadly similar, although each species feeds on a few kinds of flowers not fed on by the other species.

The foraging behavior of this group of nectar bats can be summarized as follows. Commissaris's bat is a feeding generalist that eats fruits and insects in addition to nectar and pollen. As a result, it is the most common of the four species. Underwood's bat is an uncommon nectar specialist that feeds more on uncommon flowers that produce little nectar than the other resident species. It must be a very efficient flier to be able to persist on scarce, widely spaced flowers. Of the two nonresident species, the dark nectar bat appears be a wide-ranging nomadic species whereas the larger Goldman's bat is likely to be a long-distance commuter from higher elevation day roosts. Overall, this bat community has a very different "feel" from La Selva's hummingbird community. It lacks a clear separation between understory and canopy feeders as well as the absence of territorialists that aggressively defend their food plants. Much of the bat activity here occurs in the canopy and involves uncommon, wide-ranging species.

Finally, it should be noted that some communities of nectar bats are seasonally "invaded" by opportunistic frugivorous species such as Seba's short-tailed bat, the Jamaican fruit bat, and the pale spear-nosed bat. This is particularly true in tropical dry forests during the dry season when trees such as balsa, pochota, silk cotton, and West Indian locust produce large numbers of large, very accessible flowers. Only two species of nectar bats—the common long-tongued bat and Underwood's long-tongued bat—occur in this habitat in northwestern Costa Rica. But at this time of the year substantial numbers of "invaders" can be seen landing on and drinking nectar from these nectar-rich flowers. None of them have the morphological adaptations seen in dedicated nectar bats (e.g., elongated rostrums and long tongues); they are larger than most nectar bats; and they can't hover. Nonetheless, they can feed easily at these flowers. I've caught Jamaican fruit bats in this habitat that were absolutely covered with silk cotton pollen that turns their brown fur yellow. It's not known whether these species are serious competitors with the nectar bats in this habitat. But my impression is that they aren't because of the surfeit of nectar being produced by trees at this time of the year.

CONCLUSIONS

Unlike hummingbirds, nectar-feeding phyllostomid bats have undergone a rather modest evolutionary radiation that currently contains about fifty-six species classified in two subfamilies. No family containing only nectar feeders has evolved in bats anywhere in the world. Nonetheless, the parallels resulting from convergent evolution between these birds and bats are very impressive. They include small size, the ability to hover, elongated bills or rostrums, long tongues, and biochemical machinery that quickly shunts the carbohydrate-rich nectar they ingest into energy for powering their large flight muscles. Differences seen in these two groups reflect their very different evolutionary histories. In hummingbirds these differences include a large array of brilliant plumages and fancy heads and tails, the frequent use of energy-saving torpor, and highly aggressive social systems. Their world includes a very colorful array of scentless flowers in a variety of different shapes and sizes. In nectar bats these differences include the use of echolocation and an excellent sense of smell for locating night-blooming, pale-colored flowers and a highly gregarious lifestyle. Hummingbirds have been strongly selected to be optimal foragers, but it appears that this hasn't been the case in nectar bats. Their gregarious lifestyle has often resulted in having to commute long distances from their day roosts to their feeding grounds every night. Some of them forage in groups, behavior that is never seen in hummingbirds. Once in their feeding areas, they spend a lot of time searching for and flying between flowers. They don't appear to be either time minimizers or energy maximizers in the traditional concepts of optimal foraging.

5

PUTTING THESE VERTEBRATES AND THEIR FOOD PLANTS TOGETHER

THROUGHOUT THE CENOZOIC ERA—THE past 65 Ma—vertebrate nectarivory has played a relatively minor role in the pollination biology of flowering plants compared with insects, although it often involves showy animals and flowers. In a variety of contemporary tropical habitats, for example, the frequency of bird pollination averages about 4.7 percent of species of shrubs, epiphytes, and trees. At an average of about 2.8 percent in these habitats, the frequency of bat pollination is even lower. Nonetheless, because it often involves conspicuous and ecologically important species within tropical and subtropical habitats, vertebrate pollination is still very important in these habitats.

To begin to appreciate this, picture the kinds of flowers visited by hummingbirds in a variety of different new-world habitats (figure 20). In the tropics, these include the shaving brushlike flowers of legume trees (Fabaceae) and a multitude of colorful tubular flowers produce by shrubs such as members of the coffee family (Rubiaceae), many epiphytic bromeliads (Bromeliaceae), and herbs such as mints (Lamiaceae). In moist and wet tropical forests, hermit hummingbirds are attracted to thick clumps of megaherbs of the Heliconiaceae with their large inflorescences bearing colorful bracts surrounding white or yellow tubular flowers. In the Andes, flowers produced by shrubs in the blueberry family (Ericaceae) are especially important. In subtropical and temperate North America, migrant hummingbirds visit the tubular flowers of

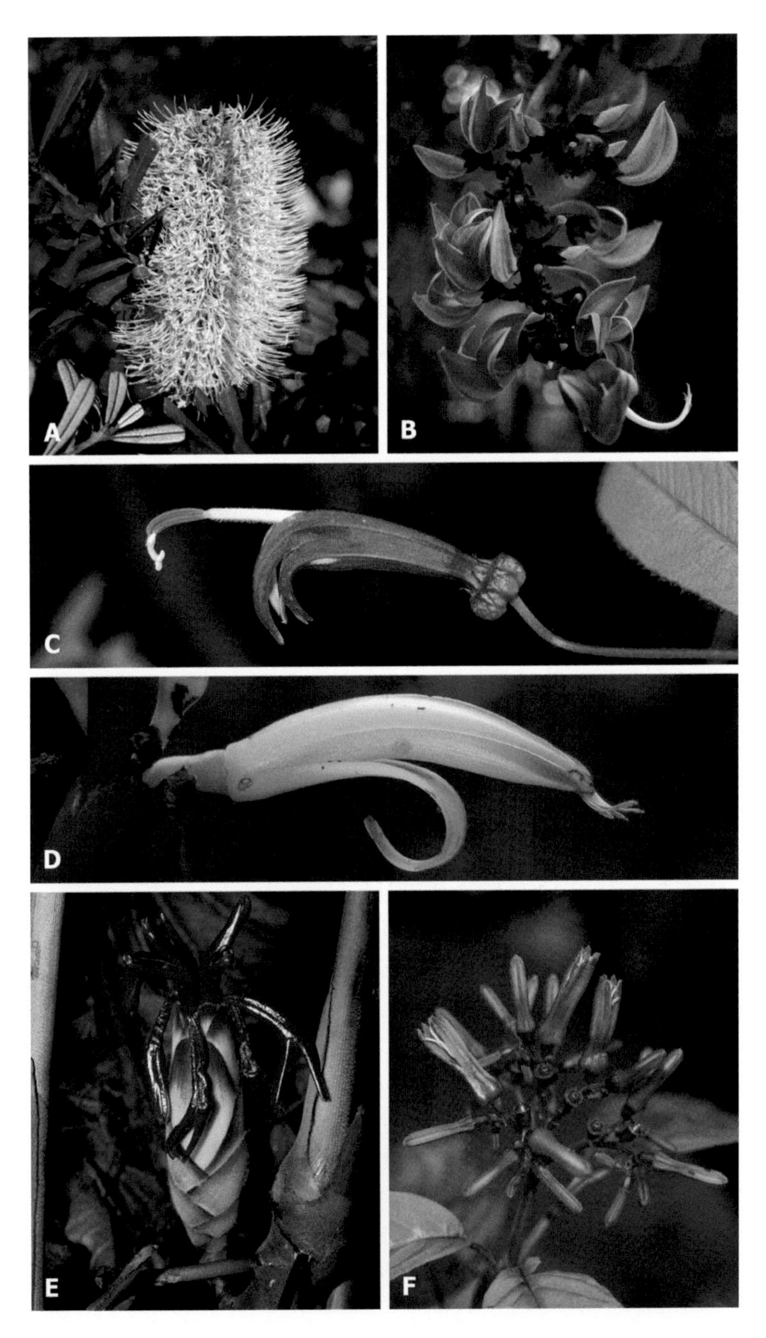

FIGURE 20 Examples of new-world (NW) and old-world (OW) nectar bird flowers: A. OW *Banksia* sp. (Proteaceae), B. OW *Butea monosperma* (Fabaceae), C. NW *Centropogon* sp. (Campanulaceae), D. NW *Heliconia vaginalis* (Heliconiaceae), E. OW *Hornstedtia scottiana* (Zingiberaceae), and F. NW *Hamelia patens* (Rubiaceae). Reprinted from Fleming and Kress (2013) with permission.

ocotillos, columnar cacti, and a host of flowers produced by herbs in the mint and foxglove families (Lamiaceae and Scrophulariaceae, respectively). In the summer, the alpine meadows of the Rockies abound with the red, orange, or purple flowers of many species of herbs that fuel the southward migrations of rufous, broad-tailed, and calliope hummers.

Most of the flowers visited by tropical and temperate hummingbirds come from epiphytes, shrubs, and herbs—plants that are mostly small in stature—but this is not the case for phyllostomid nectar bats. They certainly do visit the flowers of some epiphytes and shrubs, but the bulk of their flowers are produced by trees. In tropical habitats, trees containing bat-pollinated flowers can be found in the following families: the catalpa family (Bignoniaceae, both trees and vines), Cactaceae (columnar cacti), Fabaceae (trees), the mallow family (Malvaceae, especially trees in the Bombacoideae), and the myrtle family (Myrtaceae, trees) (figure 21). Farther north in the subtropics, these bats are important pollinators of species in two families: Asparagaceae (paniculate agaves) and Cactaceae (columnar cacti). Whereas hummingbird flowers are usually small, nectar bat flowers are often large. Examples include those of columnar cacti as well as balsa, pochota, and silk cotton trees. These trees are often more common in arid or dry habitats than in wetter habitats.

The fact that these vertebrate-pollinated flowers are in the vast minority among all angiosperm flowers might compel you to ask, why have these kinds of flowers evolved in the first place? Why have some angiosperm lineages evolved species to attract these kinds of pollinators? What are the costs and benefits of vertebrate pollination compared with the more common insect pollination? The costs to plants associated with this mutualistic interaction are obvious mainly because hummingbirds and nectar bats are much larger than most insects and are endothermic and thus have high FMRs. This means that they are generally much more costly to attract and often require larger flowers with greater amounts of nectar (and pollen) to satisfy their energetic needs. But on the positive side, the benefits of vertebrate pollination include potentially more reliable flower visitation because of their substantial cognitive abilities, their ability to carry large amounts of pollen on their fur or feathers, their often-substantial movement among plants that can result in high rates of genetic outcrossing, and their ability to remain active under harsh climatic conditions compared with some insects. Insects may remain the pollinators of choice for most angiosperms, but vertebrates also offer important advantages as pollinators.

PHYLOGENETIC DISTRIBUTION OF THESE POLLINATION MODES

Given their much greater number of species, it should not be surprising to learn that hummingbirds are known to pollinate many more flowers than phyllostomid nectar bats. Hummingbirds pollinate flowers found in about 95 families containing a total of about 500 genera and several thousand species (up to 7,000 by some accounts). In contrast, members of the much more modest radiation of phyllostomid nectar bats are known to pollinate flowers in about 44 families containing about 159 genera and 350 species.

What is the phylogenetic distribution of these families? Where in the evolutionary history of angiosperms do these families occur, and are these occurrences similar for both hummingbirds and bats? It is important to remember that the evolutionary radiation of both hummingbirds and phyllostomid nectar bats in the New World began only about twenty to twenty-two million years ago, long after the major diversification of flowering plants. So their evolutionary impact on angiosperms is relatively recent, geologically speaking. Reflecting this, most of the families containing hummingbird- or bat-pollinated flowers occur in the most advanced clades of angiosperms (see figure 1), namely in the asterids and rosids (figure 22). Although figure 22 contains more hummingbird- than bat-pollinated plant families, the trends in both families are very similar. No families of basal angiosperms contain flowers pollinated by new-world vertebrate nectar feeders. These early plants were pollinated by insects and other arthropods. Instead, vertebrate-pollinated families are concentrated in the monocots, asterids, and rosids.

Although the distribution of families containing species pollinated by hummingbirds and nectar bats are similar across angiosperm phylogeny, we can still ask to what extent do their flower choices overlap? How different are the diets of these animals at the family level? Given that the suite of flower characteristics associated with hummingbirds and nectar bats can be quite different, how often do these pollination syndromes (see box 2) occur in the same families? John Kress and I have attempted to answer these questions in broad-brush fashion by indicating that within clades, the proportion of shared families in hummingbirds and phyllostomid nectar bats ranges from 17 percent in monocots to 56 percent in rosids; values in basal eudicots and asterids are 50 and 33 percent, respectively. These results suggest that within angiosperm clades, hummingbirds and bats are visiting flowers that often occur in

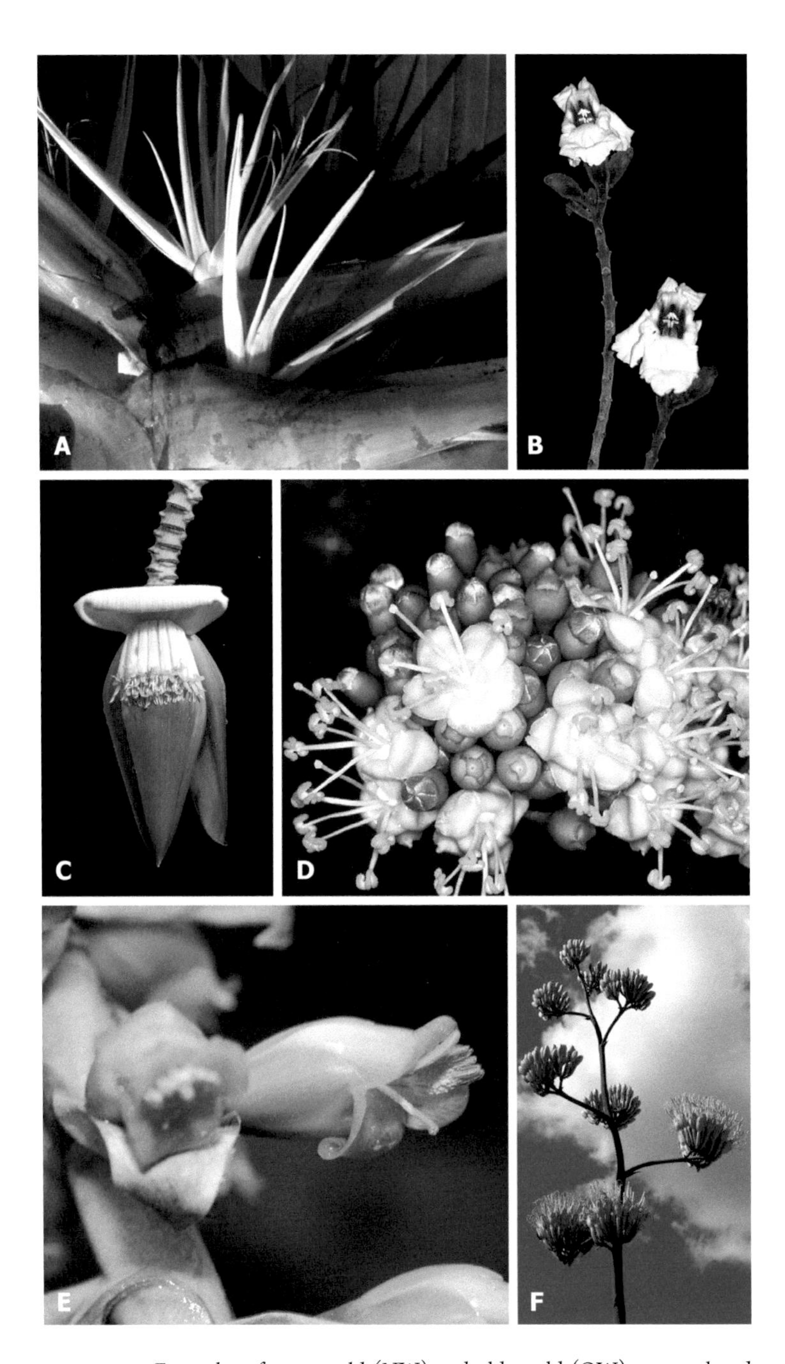

FIGURE 21 Examples of new-world (NW) and old-world (OW) mammal and nectar bat flowers: A. OW *Ravenala madagascariensis* (Strelitziaceae), B. OW *Markhamia stipulata* (Bignoniaceae), C. OW *Musa acuminata* (Musaceae), D. Worldwide *Ceiba pentandra* (Malvaceae). E. NW *Vriesia* sp. (Bromeliaceae), and F. NW *Agave chrysantha* (Asparagaceae). *Ravenala* is pollinated in Madagascar by lemurs. Reprinted from Fleming and Kress (2013) with permission.

different families. For example, within monocots, major families containing hummingbird-pollinated flowers include megaherbs in the Heliconiaceae as well as smaller herbs such as spiral gingers (Costaceae) and arrowroots (Marantaceae). Bromeliad epiphytes are important for both hummingbirds and nectar bats, but agaves (Asparagaceae) are more important for bats than hummingbirds. Likewise, in asterids, important hummingbird families include the coffee family (Rubiaceae), mints (Lamiaceae), and blueberries (Ericaceae), whereas the catalpa family (Bignoniaceae) is especially important for nectar bats. Important families for both groups within the asterids include lobeliads (Campanulaceae), Marcgraviaceae, and gloxinia relatives (Gesneriaceae). As we will see, within shared families, flowers that attract either hummingbirds or nectar bats usually occur in different genera or species. Thus, at the species level, plants in the New World have often specialized in which of the two vertebrate groups they attract as pollinators.

EVOLUTIONARY TRANSITIONS FROM INSECT TO VERTEBRATE POLLINATION

The current floral diversity in angiosperms is overwhelming—from the tiny flowers found in the two-millimeter-long aquatic plant watermeal (*Wolffia*) (Araceae) to the meter-wide flowers of Asian understory *Rafflesia* plants (Rafflesiaceae). Given an evolutionary history of at least 130 Ma and over three hundred thousand flowering plants currently, this diversity is not surprising. But what were the earliest angiosperm flowers like, and what is the general trajectory of floral evolution in angiosperms? Furthermore, what were the possible selective factors that caused angiosperms to evolve vertebrate-pollinated flowers? As an aside, the vast diversity of flowers that we see in angiosperms today is one of the wonderful aspects of the evolution of life on Earth. Botanical gardens, flower shows, and even large floral shops are excellent places to visit to begin to appreciate this diversity.

When they first evolved, it is likely that angiosperms had generalized pollination systems involving beetles and flies. These flowers were generally small and radially symmetrical (e.g., like a daisy); they initially lacked sepals (the outer green leaflike structures of the corolla that protect the flower bud) and petals (the inner, colorful structures of the corolla) but had stamens with pollen-containing anthers and seed-containing ovaries that characterize

Box 2. Pollination syndromes

The concept of pollination syndromes and whether it is useful in discussions of pollination biology has been discussed by many people for decades. Views about this have ranged from "the concept is not useful" to "of course the concept is useful." These syndromes have traditionally focused on such floral features as time of flower opening (i.e., anthesis), size and shape of flowers, flower color, floral scent (or lack thereof), amount and composition of nectar per flower, method of flower presentation, the size of flower crops, and the phenology of flower seasons. Given the substantial biological diversity of potential pollinators—from insects ranging from thrips to butterflies and nectar-feeding birds and bats—it is not surprising to find that flowers "aimed" at different subsets of these pollinators can differ greatly in their suite of floral characteristics. In 2020, Agnes Dellinger published a thorough, thoughtful, and critical review of the pollination syndrome concept, which I accept as realistic and useful.

Here I will use traditional floral features to characterize the pollination syndromes associated with hummingbirds and phyllostomid nectar bats.

ANIMAL GROUP	Hummingbirds	Nectar bats
ANTHESIS	Diurnal	Nocturnal
COLOR	Usually vivid; often red; UV patterns sometimes present	White or otherwise drab; sometimes green; UV patterns may be present
SCENT	Usually lacking	Often conspicuous and musty
SIZE AND SHAPE	Usually small; tubular with either radial or bilateral symmetry	Often large; flat "shaving brush," bell-shaped, or tubular; radial symmetry
PRESENTATION	Hanging or otherwise extending away from foliage	Hanging or otherwise extending away from foliage; sometimes on branches or trunks
NECTAR AND POLLEN	Nectar usually in small amounts; sucrose rich; little pollen	Nectar in substantial amounts; hexose rich; much pollen

These floral features are thought to reflect the morphology, physiology, sensory biology, and behavior of their major pollinators as discussed in detail throughout this book. Given that these "syndromes" are real, the Big Questions then become: Why evolve flowers to attract vertebrate pollinators, and what's involved in the transition from insect to vertebrate pollination?

nearly all angiosperms from the beginning. Nectar as a pollinator attractant was initially produced by stamens but eventually became produced by specialized glands, often at the base of the sexual organs. By the mid-Cretaceous (about 120 Ma), recent fossil evidence indicates that both generalized and more specialized pollination systems involving insects had evolved in basal angiosperms and basal monocots. These specializations included nonnectar floral rewards (e.g., protein bodies), viscin threads that bind groups of pollen grains together, and fused petals (e.g., like morning glory flowers). Bilaterally symmetrical or zygomorphic flowers that are common in many major plant groups today (e.g., in asterids and many legumes) evolved somewhat later but were widespread by the beginning of the Cenozoic era. As we will see, many but not all hummingbird and nectar bat flowers today are zygomorphic rather than radial in structure.

Botanists believe that much of the floral and taxonomic diversity that we see in today's angiosperms has been driven by coevolution between flowering plants and their pollinators. Early on, small generalist nectar- and pollen-feeding flies and beetles were attracted to flowers, but that phase of rather generalized pollination likely did not last very long, geologically speaking. By the Late Cretaceous, a vast array of flower types associated with a diverse array of angiosperm lineages (e.g., orders and families) was interacting with a much broader array of insect pollinators, including early bees and butterflies. As my friend Steve Buchmann has written (2015, p. 37): "Flowers didn't just get larger, they became brighter, showier, and more fragrant. Bigger, more colorful petals helped make flowers more visible against the green background of leaves, better able to attract and hold the attention of pollinators. . . . Bigger blossoms also contained more stuff—greater numbers of anthers and therefore greater quantities of pollen, as well as nectary tissues producing more of the nectar needed by flower-visiting insects." It is against this background that the stage was eventually set for the evolution of nectar-feeding hummingbirds and nectar- and pollen-feeding bats, which are relative newcomers on the pollination scene.

HOW TO MAKE A FLOWER

Before discussing the evolution and coevolution of flowers and their vertebrate visitors in detail, let's first see how flowers generally evolve—that is, what are the genetic and developmental mechanisms involved in the transformations that occur from one floral morphology to another? To begin with, it should be noted that flowers are composed of several organs just as the skulls, wings, hearts, and brains are animal organs (sensu lato). Therefore, they have a modular structure based on a complex set of genetic and developmental actions and interactions. As schoolkids we all learned the basic structure of typical angiosperm flowers (i.e., petals and their sexual parts). But, not surprisingly, professional botanists identify a whole host of additional features associated with angiosperm flowers. Some of these include the spatial arrangement of sexual parts within flowers (phyllotaxis), the number of different floral parts (merism), floral symmetry, fusion of petals into floral tubes (sympetaly), and many others. The bottom line here is that flowers are complex organ systems whose evolution has involved a number of different pathways. Two of these include (1) the use of existing structures for new functions (e.g., the conversion of stamens into nectar guides or new petals) and (2) the fusion of plant parts to form new organs (e.g., the creation of pollen packets [pollinia] in orchids and milkweeds). As Peter Endress, a major student of flower evolution, has indicated, processes such as convergence or the repeated evolution of similar floral structures in unrelated lineages has been rampant during floral evolution.

Modern studies of the evolution and diversification of flowers are based on an ever-expanding set of tools including classical morphological studies of fossil and contemporary plants, genetic and developmental studies, statistical analyses, and computer modeling. It is a large and complex field of research of which I will barely scratch the surface here. Nonetheless, I feel that some generalities that have emerged from these studies are useful for understanding some of the complexity of this subject. At its most general level, for example, Louis Ronse de Craene of the Royal Botanic Garden of Edinburgh has stated (2018, p. 387): "Floral evolution is the result of subtle developmental changes leading to significant changes in the morphology of flowers. The historical context [i.e., genetic changes] explains the diversity at a given stage in time, but influences of physical forces operate before, during, and after the ontogeny of the flowers." That is, the interactions of genes and developmental processes alone cannot fully explain the evolutionary diversification of flowers; physical processes such as those associated with spacing within the developing flower are also involved. He further states (p. 369) that morphological changes "can be triggered by different causes such as genetic mutations, pollinator-mediated selection on flower size and shape, and mechanical forces acting directly on the position, shape, and size of organs."

With this general background, we can now look a bit more deeply into the evolution of flower morphology before moving on to a closer examination of the coevolution between hummingbirds and phyllostomid nectar bats and their flowers. We've already seen many of the adaptations on the animal side of this interaction, so let's look at the plant side. To begin with, flowers are highly modified leaves that develop at the tips of branches (i.e., in the apical meristem). Their basic structure includes whorls of organs beginning with the outermost sepals followed by inner whorls of petals, male parts (the androecium and its pollen), and the innermost female parts (the gynoecium and its ovules). From this basic structure evolution has created all kinds of variations, sometimes involving a few and sometimes involving all of these basic structures. For example, tubular flowers with zygomorphic symmetry result from the fusion of petals that can vary widely in length and width (picture the large tubular flowers of jimsonweed, a hawkmoth flower). As we know, tubular flowers are a hallmark of many hummingbird and nectar bat flowers. Compared with other floral parts, the number of anther-bearing stamens can be highly variable (picture the stamen-rich flowers of columnar cacti such as saguaros). Like petals, they can also be fused together into a single structure in some families. The gynoecium containing one or more carpels, which are sometimes fused, and their ovules is unique to angiosperms. Deposition of pollen on the surface of the stigma, a vertical extension of the carpel, rather than directly on ovules as in gymnosperms, is another angiosperm innovation.

Nectaries are glandular tissues that produce the main nutritional reward for most pollinators. They occur in many kinds of plants, including gymnosperms, and are not necessarily located only in flowers. When they are located outside of flowers, they often attract ants that are thought to protect plants from herbivores. When they are located in flowers, they usually occur at the floral base between the androecium and the gynoecium; they can also occur within the ovary(ies) in some groups (e.g., in certain monocots).

FLORAL DEVELOPMENTAL GENETICS

In 2018, Douglas Soltis and colleagues wrote (p. 367): "The most obvious shifts in floral morphology seem to be controlled by variation in expression of transcription factors, which may then trigger complex downstream cascades of genes that encode more specific attributes of floral features." Transcription factors are proteins that bind to specific sequences of DNA and turn them on or off together. From this summary statement, the following account of floral development summarizes in simplified form the discussion of floral developmental genetics from their book.

The major organs found in flowers (e.g., sepals, petals, etc.) can be described by the so-called ABCE model of organ development in which A, B, C, and E refer to four major sets of genes that act separately or together to produce floral organs. Thus, A genes specify sepal formation; A and B genes together specify petal formation; B and C genes together specify stamen formation; C genes specify carpel formation; and E genes contribute to the formation of all of the organs. These sets of genes are now called MADS-box genes because of their fundamental role in the development of particular floral organs. They operate in a similar fashion to HOX genes in animals (see page 29), indicating that broadly similar genetic pathways of morphology or organ development have occurred during the evolution of both plants and animals. In flowering plants, specific genes within these sets have been identified and have been given interesting names such as APETALA1, PISTILLATA, and AGAMOUS in the A, B, and C sets, respectively. Although the ABCE model is widely applicable to eudicot (advanced) angiosperms, it is less applicable to basal angiosperms for reasons we don't need to be concerned with here.

Floral symmetry—either radial (actinomorphic, picture a sunflower) or bilateral (zygomorphic, picture a snapdragon)—has long been a defining feature of different groups of angiosperms. Radial symmetry clearly is the ancestral or basal condition and bilateral symmetry is a derived (advanced) condition in angiosperms. The developmental genetics of floral symmetry has therefore been of great interest to students of angiosperm evolution. Working with snapdragons, which are zygomorphic in the advanced rosid lineage (figure 1), researchers have identified three sets of genes containing development initiation transcription factors—CYCLOIDEA (CYC), DICHOTOMA (DICH), and DIVARICATA (DIV)—that are responsible for producing this flower's zygomorphic morphology. Zygomorphy has evolved independently in many angiosperm lineages, and it is known that CYC-like genes are involved in many, if not most, of these cases. In some lineages, actinomorphic flowers have evolved from zygomorphic ones (i.e., an evolutionary reversal), and it is likely that loss or deactivation of the CYC genes has occurred in these cases.

We've seen that the skulls of hummingbirds and nectar bats are composed of modular units that evolve as genetically integrated units. So it should come as no surprise that a similar situation occurs in the evolution of angiosperm flowers. Indeed, recent work with many different flowering plants has shown that the major organ systems controlled by the ABCE factors can evolve either separately or in concert with each other. Given this situation, it is easy to understand why there are myriad ways in which different flower morphologies can evolve. Often in response to selection by different kinds of pollinators (or horticulturalists), it is known that flower morphologies within families, genera, and even within species can evolve quickly (i.e., within a few million years or much less). For example, during its domestication (i.e., for about five thousand years), flowers of roses have become substantially larger and their number of petals has increased from an ancestral five to forty or more. Petal number has increased via the conversion of some of its stamens into petals. Importantly, many of the genes involved in the evolution of different aspects of flower morphology are now known in certain model species (e.g., *Arabidopsis,* a rosid in the mustard family [Brassicaceae]). I discuss examples of some of the transitions involved in the evolution of hummingbird and nectar bat flowers below.

OTHER FLORAL TRAITS

In addition to flower size and shape, a variety of other flower traits, including flower color, nectar composition and scent, and method of presentation are conspicuous features of pollination syndromes. John Kress and I have reviewed these traits in detail, and I will only provide a summary of our results here.

FLOWER COLOR

Bee-pollinated flowers, which are ancestral to many vertebrate flowers, are typically yellow, blue, or purple whereas those of hummingbirds are often red or orange; bat-pollinated flowers tend to be white or otherwise light colored. Major pigments producing the colors found in flower corollas include anthocyanins that produce orange-red to purple colors, flavanols that produce yellows, and carotenoids that produce yellows and oranges. Of these, the genetics behind the complex biosynthetic pathway producing anthocyanins is best known. In Temperate Zone flowers, the transition from blue insect-pollinated flowers to red hummingbird-pollinated flowers is asymmetric in the sense that the blue-to-red transition occurs more often than the reverse transition. A similar bias is known to occur in the transition from pigmented to white flowers pollinated by bats. This transition often involves loss-of-function mutations that occur in the anthocyanin biosynthetic pathway.

NECTAR

This sugar-rich aqueous solution is the main energy reward for animal pollinators. It contains about 90 percent sugars by weight. These include the monosaccharides (hexoses) glucose and fructose and the disaccharide sucrose. The remaining 10 percent includes a mixture of amino acids, lipids, minerals, and secondary compounds (herbivore deterrents). As you might guess, the amount of nectar produced per flower is positively correlated with flower size and its composition varies with type of pollinator. Many small hummingbird flowers produce only microliters of nectar per day compared with some large nectar bat flowers that produce up to 15 milliliters of nectar per day. Sugar concentration in these nectars for both kinds of pollinators averages about 20–30 percent and is lower than typical bee-pollinated flowers (about 45 percent). Many studies have shown that the nectar of hummingbird flowers is dominated by sucrose whereas the nectar of phyllostomid nectar bat flowers is dominated by the hexose sugars glucose and fructose. Interestingly, these differences tend to hold at the family, genus, and species levels. That is, species within the same genus that are pollinated either by hummingbirds or bats differ in the kinds of sugar in their nectar, which suggests that sugar composition in nectar is evolutionarily very labile and is responsive to selection by different kinds of pollinators.

It is not yet clear, however, why the nectar in hummingbird flowers tends to be sucrose-rich whereas the nectar in nectar bat flowers is hexose-rich. Experimental work with hummingbirds and glossophagine bats has shown that they both prefer to ingest sucrose-rich nectar rather than hexose-rich nectar. These preferences apparently are not related to differences in the digestibility of sucrose and hexoses. Not all glossophagines, however, have a sucrose sweet tooth. The common long-tongued and lesser long-nosed bats are equally attracted to sucrose- and hexose-rich solutions of equal caloric value. These results therefore raise a question about the extent to which phyllostomid bats have been strong selective agents for the sugar composition of their flowers.

FLORAL SCENTS

Flowers often produce a rich bouquet of floral scents composed of a complex mixture of many volatile compounds of low molecular weight. These include aliphatics, benzenoids, terpenoids, nitrogen-containing compounds, and sulfur-containing compounds. The genetics and biochemistry are known for a number of them. Scents are an ancient way in which plants, even those predating the evolution of angiosperms, have communicated with animals, most notably insects. Rob Raguso, a major student of floral scent production, has stated that scents can often be as important as visual cues for attracting pollinators. Additional functions of these compounds include repelling herbivorous insects and arresting the growth of microbes.

As we've seen, hummingbirds and bats differ in their olfactory senses, and their flowers reflect this. Most hummingbird flowers produce no (or very little) scent, whereas bat flowers produce a rich array of scents. In the neotropics but not in the paleotropics, for example, bat flowers often have scents dominated by sulfur compounds (for the curious, these include dimethyl disulphide and 2, 4-dithiapentane) that are really stinky. These kinds of compounds are not commonly used as scents by most other kinds of flowers. Experimental work with the common long-tongued bat indicates that they apparently have an instinctual preference for sulfur-rich compounds. Evidence supporting the idea that floral scent production is under strong selection by pollinators comes from comparisons of closely related flower species. Hummingbird-pollinated species typically lack scents, whereas bat-pollinated flowers have them. In the world of pollinators, therefore, hummingbirds are the odd men out. Most invertebrate and vertebrate pollinators, including us, are strongly attracted to scented flowers, but hummers are not. Imagine what a flower show or flower shop would smell like if they were stocked only with hummingbird flowers.

FLOWER ACOUSTIC FEATURES

In addition to scent, some tropical flowers have taken advantage of another sensory adaptation that is lacking in hummingbirds to attract nectar bats. As we've seen, new-world nectar bats are echolocators whereas their old-world counterparts are not (see below). Auditory cues are important for orienting toward cactus flowers (and likely, hummingbird feeders) in lesser long-nosed bats. And the same is likely to be true for other glossophagine and lonchophylline bats. Field observations and experimental work with the flowers of two tropical vines—*Mucuna holtonii* (Fabaceae) and *Marcgravia evenia* (Marcgraviaceae)—indicate that a triangular floral projection in *Mucuna* flowers or dish-shaped leaves above *Marcgravia* inflorescences serve as acoustic mirrors that bats use to locate flowers. Old-world relatives of these flowers do not have these kinds of modified flowers or leaves, indicating that they have evolved specifically to attract echolocating bats.

Another way that flowers can increase their visibility to echolocating bats is to be surrounded by ultrasonic absorbing (rather than ultrasonic reflecting) material. The flowers of many bat-pollinated cacti, for instance, are located on branches in the midst of *cephalia*—zones of dense "hairs." Recent experimental work with an Andean columnar cactus (*Espostoa frutescens*) has shown that these zones absorb, rather than reflect, ultrasonic sounds in the frequencies produced by echolocating nectar bats. Flowers thus stand out from background acoustic reflections, making it easier for bats to detect them. When placed on branches away from cephalia, flowers do not stand out acoustically and hence are not as "visible" to bats as they are when they occur in the cephalium.

FLOWER PRESENTATION

Plants present their flowers to pollinators in many different ways (figures 20 and 21). For example, flowers can occur singly or grouped into inflorescences; they can be presented close to or away from foliage; and they can include strong perches or not. Hummingbirds and nectar bats hover when visiting flowers, so their flowers are often presented singly rather than in flower-rich inflorescences; when they do occur in inflorescences, as in paniculate agaves, only one or a few flowers are usually open each day (or night). Most flowers visited by these animals are also presented away from foliage, sometimes in dramatic fashion on long vertical stalks or on the ends of long peduncles or leafy stems (a condition known as *flagelliflory*). A few bat-pollinated flowers such as those of the calabash tree (Bignoniaceae) are produced directly on branches or trunks (a condition known as *cauliflory*). Whatever the method of presentation, it is clear that plants have often been selected to make their flowers as accessible to their pollinators as possible.

WHEN TO FLOWER—FLOWERING PHENOLOGY

Species of plants do not exist in an ecological vacuum. Instead, they exist in plant-rich habitats full of other species that are constantly attempting to attract pollinators. To avoid serious competition for pollinators, plants have tended to specialize on particular subsets of pollinators through their many floral characteristics, including their daily and seasonal flowering times, thereby creating specific pollination syndromes (box 2). It is important to realize, however, that these syndromes are not inviolate. Pollinators do not always recognize man-made classification systems and often visit flowers that they're not supposed to as they forage for nectar. Thus, many hummingbirds, especially those with short bills such as Coquettes and Bees, frequently visit flowers in insect pollination "syndromes." As a result, these birds often have a wider array of flowers to visit than long-billed hummers such as Hermits and Brilliants that tend to restrict their visits to long-corolla flowers. Because of their larger size and nocturnal lifestyle, bats are less likely to visit diurnally opening flowers that occur in different pollination syndromes than are hummingbirds.

Assuming that pollination syndromes are real, plants within them are still likely to compete for the same kinds of pollinators as I discuss below. But competition for pollinators is not the only kind of interspecific interaction flowering plants are engaged in. It is possible, as some have pointed out, that different plants attracting the same pollinators might actually benefit from having overlapping rather than temporally separated flowering times. This is because overlapping flowering times can increase the overall amount of nectar available to pollinators and can thereby support larger pollinator population sizes. Whenever this happens, flowering plants become mutualists (i.e., they all benefit from the situation) rather than strict competitors (i.e., antagonists).

In strong contrast to the tropics, for instance, consider the situation in the montane meadows at 2,886 m elevation on the eastern slope of the Rocky Mountains in Colorado. Here the growing season is about three months long (mid-June to mid-September) so there is much less time for hummingbird- and insect-pollinated plants to evolve totally separate flowering seasons. As a result, temporal overlap in flowering times is common here, which leads to higher nectar levels for pollinators. As you might expect, flowering overlap tends to decrease in years of early snowmelt compared with years of late snowmelt in which the length of the growing season is shortened. Research on the arrival times of migratory broad-tailed hummingbirds in this area suggests that with global warming and earlier snowmelt, a mismatch could develop between when their food plants flower and when the birds arrive to breed.

Whenever plants compete for pollinators, one way they can begin to reduce this competition is to flower at different times of the day or year than other species. That is, whenever environmental conditions permit, we might expect to see the evolution of nonrandom flowering times among plants that live together and potentially compete for the same pollinators. A classic example of this is Gary Stiles's research on the flowering times of hermit-pollinated

Heliconias and other plants at La Selva (figure 23). At this lowland rainforest site, peak flowering times of ten important hummingbird food plants are scattered throughout the year in a seemingly nonrandom pattern. Similar staggered flowering seasons are also known to occur in several trees and shrubs pollinated by bats in Costa Rican tropical dry forest and in four species of bat-pollinated epiphytic bromeliads in the lowland tropical forests of Veracruz, Mexico.

Although interspecific interactions such as mutualism and competition likely play an important role in determining when particular vertebrate-pollinated plants should flower, it is important to remember that seasonal fluctuations in temperature and rainfall ultimately play the most important role in the evolution of flowering times and patterns. Tropical habitats where most hummingbirds and nectar bats live obviously have longer growing seasons than temperate, especially montane, habitats. Hence tropical plants should have more "choices" regarding when to flower than temperate plants. But even in the tropics, seasonal patterns of rainfall often occur in most habitats, and rainfall seasonality is known to have a strong effect on the flowering times of

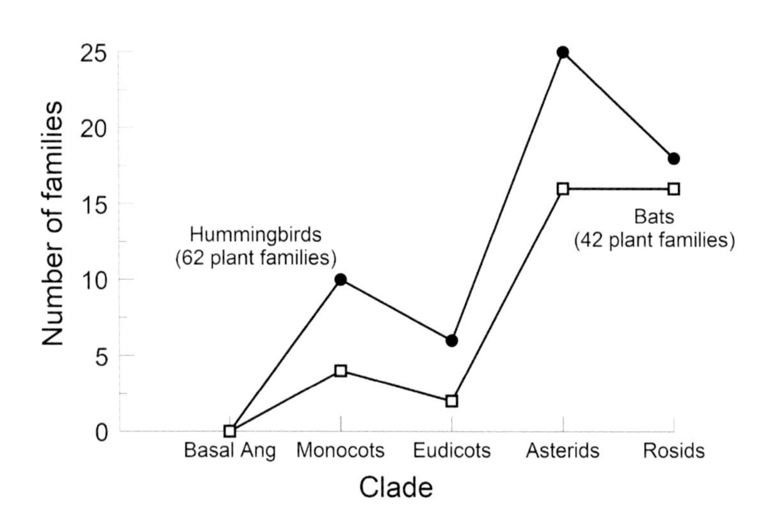

FIGURE 22 The number of families in the major lineages of flowering plants (see figure 1) containing species pollinated by hummingbirds or nectar bats. Redrawn from Fleming and Kress (2013).

tropical plants. Plants living in tropical dry forest, for example, experience very strong rainfall seasonality and as a result, many vertebrate-pollinated plants flower in the dry season so that they can produce their fruits and seeds in the wet season when conditions for seed germination are favorable. Rainfall seasonality is usually less pronounced in tropical wet forests, but seasonal flower production is often the rule there, too. For instance, Gary Stiles's work at La Selva indicates that peak flowering in hummingbird-pollinated plants occurs in the late dry and early wet seasons (February–March) whereas lowest flower numbers (and nectar levels) occur in the wet months of November and December.

Further complications involved in the evolution of flowering times in hummingbird- and bat-pollinated plants are associated with plant growth form (e.g., canopy trees versus understory shrubs) and habitat type (e.g., primary forests versus secondary forests). But we don't need to dwell on these complications here. John Kress and I have provided a comprehensive discussion of all of these factors in *The Ornaments of Life*. The bottom line here is that the evolution of flowering times in these plants can be complicated. This is because we all live on a climatically variable planet in which evolution of life's features always takes place in environments that vary in space and time. Earth's enormous biodiversity is the result of this environmental spatiotemporal variability. If this variability didn't exist, we would live on a rather dull planet indeed.

GENETIC CONSEQUENCES OF VERTEBRATE POLLINATION

When hummingbirds and nectar bats visit flowers, they acquire pollen grains on their feathers and fur. Thus, the faces and beaks of hummingbirds are often colored by the pollen they've acquired. A similar situation occurs in nectar bats whose snouts and faces often become pollen-covered. Experiments have shown that bat fur usually picks up more pollen per flower visit than hummingbird feathers. And they are likely to deposit more of that pollen on the stigma(s) of several of the next flowers they visit. Since many of the plants they visit have self-incompatible breeding systems in which flowers need to receive pollen from another plant of the same species for fertilization of their ovules, plants are under selective pressure to cause their pollinators to move to another plant (hopefully of the same species) to avoid wasting pollen on unproductive visits to flowers of the same plant (or different species).

One way they do this is by limiting the amount of nectar they produce in their flowers, forcing pollinators to move on for more food. But pollinators are not always faithful visitors to other individuals of the flowers they've just visited. After all, high-energy pollinators such as hummingbirds and nectar bats are probably under strong selection to maximize their rates of energy gain while foraging. So it's not surprising to know that they can be promiscuous regarding their flower choices. Many studies have shown that both hummingbirds and nectar bats can be carrying the pollen of as many as six different flower species on their fur or feathers at one point in time. In Costa Rican tropical dry forest, for instance, we found that 16–79 percent of the seven species of phyllostomid bats we examined were carrying mixed-species pollen loads on their fur when they were captured.

Does this apparent flower choice promiscuity have any important consequences for plants? One obvious consequence is that pollen deposited on the stigmas of nonconspecific flowers is "wasted," i.e., it cannot be used to pollinate flowers of the "correct" plant species. This should have at least two evolutionary consequences. First, it should select for greater pollen production per flower to compensate for pollen lost via deposition on the wrong flowers. And second, it might be expected to increase selection for nonoverlapping flowering times for plants attracting the same pollinators. Other potential "costs" to plants because of their promiscuous pollinators include reduced fruit set and seed production in flowers that receive the "wrong" pollen. Empirical studies with hummingbird-pollinated plants, however, indicate that this doesn't necessarily always happen.

My experimental work in the Sonoran Desert with fruit-set in organ pipe columnar cacti has produced surprising results regarding the consequences of receiving the "wrong" pollen. In April and early May, early flowering organ pipe plants are definitely in competition for visits by nectar bats with two other flowering columnar cacti—cardón and saguaro. Cardón is an especially potent competitor because its flowers are about fifty times more abundant than those of organ pipe at this time of year. So what happens when cardón pollen is experimentally placed on the stigmas of organ pipe flowers, which must occur frequently at this time of year? Based on the results of lots of other studies, we would expect that organ pipe flowers receiving only the "wrong" pollen will abort. But this doesn't happen. Instead, flowers receiving cardón pollen are as likely to produce mature fruit as control flowers receiving only organ pipe pollen. The only problem is that fruits produced from cardón pollen contain seeds that lack embryos. They are sterile, and their production should be strongly selected against. Incidentally, as expected, cardón flowers receiving only organ pipe pollen always abort. Only organ pipe can produce fruit with the wrong pollen at this study site. At other places in Sonora where cardón is absent, however, early flowering in organ pipe is not likely to be selected against because it occurs when nectar bats and hummingbirds are migrating through the area. So geographic location matters for early flowering organ pipe plants.

Ultimately, animal pollinators provide mobility for the pollen their food plants produce, and this can result in gene flow within and between populations of these plants. But not all pollinators produce similar amounts of gene flow for their plants. Some species provide only short-distance gene flow whereas others provide a mixture of short- and long-distance gene flow. Gene flow in plants and animals is important for at least two reasons. First, it determines a species' effective population size or genetic neighborhood, i.e., the number of individuals in a population that can actually or potentially share each other's genes during reproduction. And second, it determines a species' speciation potential. In the absence of significant levels of gene dispersal or gene flow between populations, species with small genetic neighborhoods are more likely to diverge genetically from each other as a result of random events or strong selection than populations with large genetic neighborhoods. Genetic divergence between populations is an early stage in the formation of new species. In the presence of significant levels of gene flow, however, genetic divergence within both small and large effective populations can slow down or stop. Theoretical studies have shown that it takes only a small amount of interpopulation gene flow per generation to prevent populations from diverging genetically.

Where do hummingbirds and bats fall along this pollen and gene mobility continuum? As you might guess, most hummingbirds are likely to move pollen and genes shorter distances within and between populations than nectar bats. And within hummingbirds, territorial species, which includes most kinds, are likely to move pollen and genes shorter distances than trapliners such as Hermits. Genetic studies have shown that territorial hummingbirds do indeed usually move pollen and genes shorter distances (e.g., < 20 m) than trapliners (e.g., ≥ 180 m). Territorial intruders such as females and behaviorally subordinate species, however, are likely to move pollen and genes greater distances than territorial males. So even in species pollinated primarily by territorial males gene flow can sometimes be extensive.

Compared with hummingbirds, phyllostomid nectar bats sometimes move pollen and genes several kilometers in a night. They clearly are longer-distance pollinators than most of their avian counterparts. And most importantly in a world in which forests and other habitats are becoming fragmented, these bats likely play a critical role in keeping isolated plants and populations connected genetically. For example, in tropical dry forest in Mexico and Costa Rica, certain nectar bats, including common long-tongued and lesser long-nosed bats, are important pollinators of flowering trees such as kapok in both intact forest and forest fragments. Similarly, nectar bats are known to pollinate flowers of the West Indian locust tree in fragmented populations of tropical dry forest in Puerto Rico. Because of this, fragmented populations are not likely to become genetically depauperate as quickly as theory might suggest.

What are the speciation consequences of these differences in pollen and gene movement? Are speciation rates higher or lower in vertebrate-pollinated plants than relatives that are not vertebrate pollinated? And are speciation rates higher in hummingbird-pollinated plants than in nectar bat pollinated ones? John Kress and I have reviewed data that bear on the second question and note that many authors have reported that hummingbird-pollinated groups of plants often have more species and hence higher speciation rates than their insect-pollinated relatives. These differences have been found in many kinds of neotropical plants, including epiphytes (e.g., in the pineapple family Bromeliaceae), shrubs (e.g., in acanths, Acanthaceae), trees (e.g., in the legume family Fabaceae), and herbs (gingers in the Costaceae). Speciation rates are also notably higher in either bird- or bat-pollinated than in insect-pollinated lineages in the cactus family Cactaceae. Regarding the third question, my guess is that speciation rates in hummingbird-pollinated plants are significantly higher than those of bat-pollinated plants although few data are yet available to test this hypothesis. But the substantial difference in the species richness of hummingbird-pollinated epiphytic bromeliads compared with bat-pollinated ones (i.e., fifty-three vs. four species) suggest that this hypothesis is likely to be correct. A similar disparity exists in a clade of high-Andean passionflower vines (*Passiflora*) in which fifty-six species are hummingbird pollinated compared with only seven bat-pollinated species. My overall conclusion here is that nectar-feeding birds and bats certainly do have a significant evolutionary effect on speciation in their food plants but that this effect is likely to be complex.

WHAT IS COEVOLUTION?

Since the word *coevolution* occurs in the title of this book, it is important that we understand what this word means. Its literal meaning, of course, signifies "evolving together." But what does this word mean in modern evolutionary studies? Prior to 1980, this term was not rigorously defined and had no generally agreed-upon meaning. In that year, the tropical ecologist Daniel Janzen suggested that coevolution between two species occurs when a trait of one species evolves in response to a trait of another species, which has responded to a trait of the other species. But it was quickly recognized that this definition is really too restrictive because it ignores the reality that groups of organisms (e.g., hummingbirds and their flowers), rather than pairs of species, usually respond to each other's traits. So coevolution has been redefined more broadly as reciprocal evolutionary interactions between groups of organisms. These groups can sometimes include single lineages interacting with each other or they can include several unrelated groups of organisms interacting with another similarly unrelated group of organisms. An excellent example of the latter situation can be found in pollination syndromes (box 2) in which the flowers of many lineages of plants have converged morphologically to attract and be pollinated by particular groups of animals.

While this broader definition of coevolution may seem to be rather "diffuse" because it often involves nonspecialized (i.e., non-one-on-one) relationships between groups of pollinators and their food plants, this is not always the case. Specialized (or taxonomically restricted) plant-pollinator relationships have certainly evolved in the neotropics and elsewhere. In hummingbirds, for example, these include interactions between *Heliconia* megaherbs and hermit hummingbirds and between Andean flowers of the blueberry family (Ericaceae) and Coquette hummingbirds. In phyllostomid nectar bats, examples include several species of glossophagines and flowers of the catalpa family (Bignoniaceae) and *Leptonycteris* bats and flowers of columnar cacti (Cactaceae) and paniculate agaves (Asparagaceae).

Stefan Abrahamcyck and colleagues have examined the degree of coevolution between several well-known sets of hummingbirds and their food plants. Examples include Anna's hummingbird and gooseberry plants (genus *Ribes*) in southern California and sicklebill hummingbirds and *Centropogon* flowers in the Andes. Expected results of this coevolution include close matches between the morphology, geographic distributions, and evolutionary ages of these partners. In a set of five hummingbird/plant groups that included straight-billed to sickle-billed species, morphological matches between bill length and curvature and corolla lengths and shapes are very high. This is what we would expect since this match facilitates efficient removal of nectar. Close matches between evolutionary ages and geographic distributions of the five sets of plants and birds, however, were less. Species in only two of the five sets were similar in evolutionary age; birds were significantly older than plants in the other three sets. Similarly, species in only three of the five sets had similar geographic ranges with birds in the other two sets having larger ranges than their food plants. In another study, these researchers found a close match between bill and corolla lengths and evolutionary ages of the sword-billed hummingbird and species of passion vines (*Passiflora*) in the Andes. Based on the results of both of these studies, they concluded that coevolutionary processes are often highly flexible and dynamic and don't necessarily have to be totally congruent in time and space. Morphological matches between plants and their pollinators are particularly dynamic and can probably evolve relatively rapidly (i.e., in a few million years or less).

PUTTING IT ALL TOGETHER AGAIN: CASE STUDIES OF THE EVOLUTION OF HUMMINGBIRDS, NECTAR BATS, AND THEIR FLOWERS

With the appearance of detailed phylogenetic histories of many groups of flowering plants based primarily on DNA data, it has become possible to overlay the pollination biology of different species on these histories to understand how different pollination systems have evolved. Using this method, questions we can ask include (1) what are the ancestral pollination states in lineages containing bird- and bat-pollinated species; (2) how quickly can plants change from one kind of pollinator to another kind; (3) what features of plants and their flowers do these changes involve; and (4) are these changes reversible? My intent here is to review several studies that address these kinds of questions. I've chosen them to cover much of the geographic range of new-world hummingbirds, nectar bats, and their flowers.

Let's first consider the evolutionary history of herbs of the very diverse genus *Ruellia* (about 275 species in the Acanthaceae, an asterid) that are often called wild petunias. Members of this genus occur in many different habitats in the New World. Most of their flowers are brightly colored from the production of purple or red anthocyanins. Different species of this genus are pollinated by bees, butterflies, hawkmoths, hummingbirds, and nectar bats. Bee-pollinated flowers are purple and have broad petals that serve as landing platforms whereas hummingbird flowers are red and tubular; both of these flower types open during the day. In contrast, hawkmoth- and bat-pollinated flowers are nocturnal with the former flowers being white with a very long narrow corolla tube and the latter flowers being pale yellow or green with wide corollas and strongly exserted anthers and stigmas. Given these differences, we can ask what was the basal or ancestral pollinator in these plants, how many changes from the ancestral pollinators to other pollinators have taken place, what morphological changes in flowers have occurred during these changes, and can these changes be reversed?

Erin Tripp and Paul Manos have addressed these questions using a species-level phylogeny of *Ruellia* based on 116 species in addition to detailed analyses of floral morphology and observations of pollinators in the field. Their results indicate that purple and red flowers are most common in this genus while white or yellow flowers are uncommon. Overall, bee pollination is likely to be ancestral in the genus, but within different clades, either bee or hummingbird pollination is ancestral; hawkmoth and bat pollination are always derived conditions with bees and hummingbirds being their closest ancestors, respectively. As expected, morphological characteristics associated with different flower colors and pollinators differ in 3D morphological space, nicely reflecting the evolution of features associated with the different pollination syndromes. Moth and bat pollination appear to be evolutionary dead ends, i.e., they don't give rise to the other pollination types. In contrast, evolutionary reversals from hummingbird back to bee pollination are common. Overall, the evolution of flower morphology and its effect on major pollinators is highly labile in this genus. Because it is a young genus that is strongly associated with recent uplift of the northern Andes, morphological changes from one kind of pollinator to another probably have occurred relatively quickly (i.e., in << 5 Ma) and likely reflect responses to differences in the abundance of different pollinators in particular habitats. For example, the high diversity and abundance of hummingbirds in the Andes has resulted in the evolution of many hummingbird-pollinated species from their bee-pollinated ancestors, whereas the low diversity and abundance of nectar bats throughout the neotropics has limited the evolution of bat pollination in *Ruellia*. Finally, though uncommon, bat pollination has evolved from both bee- and hummingbird-pollinated ancestors.

Let's now turn our attention to another plant group that is common in the Andes: neotropical bellflowers (family Campanulaceae, subfamily Lobelioideae, also an asterid) of the genera *Centropogon*, *Siphocampylus*, and *Burmeistera*. Like *Ruellia*, this group is geologically very young, and its flowers are pollinated only by hummingbirds and nectar bats. Their flowers are tubular but differ in many aspects associated with either hummingbird or bat pollination. Thus, hummer flowers are usually red in color, open diurnally, lack a scent, and have long, narrow corollas. Bat flowers are dull colored, open at night, are strongly scented, and have shorter but wider corollas; they also have larger anthers containing more pollen. A principal components analysis (PCA) confirms that the morphologies of hummingbird and nectar bat flowers differ in many of their features.

Modern statistical methods allow us to infer many aspects of the evolutionary history of this group of plants (and many others). For example, in the case of bellflowers, pollination by straight-billed hummingbirds (e.g., Emeralds and Brilliants) is the ancestral condition and evolution of flowers pollinated by very curve-billed hummers (of the genus *Eutoxeres*, a hermit) and nectar bats are derived conditions. Bat pollination has evolved independently from straight-billed hummer flowers about thirteen times and the reverse situation has evolved about eleven times. Further statistical analysis indicates that the morphological features associated with the two major pollination syndromes (hummingbirds and nectar bats) likely evolved in correlated (rather than random) fashion, implying that these features have a correlated underlying genetic basis. Additional analyses suggest that the evolutionary transition from hummingbird-pollinated flowers to nectar bat–pollinated flowers has been about 5.6 times faster than the reverse transition. Finally, somewhat surprisingly, overall speciation rates in this group of plants has been similar in hummingbird- and bat-pollinated species (about 1.8 speciation events per Ma per lineage), suggesting that factors other than pollinator specialization such as habitat affinities are involved in overall diversification rates. In summary, as in the case of *Ruellia* flowers, the results for Andean bellflowers support the hypothesis that the evolution of different pollination syndromes involving vertebrates is very dynamic and has likely occurred over short periods of time.

Further support for the validity of the pollination syndrome concept comes from an analysis of the pollination biology of 23 species in West Indian plants in the asterid family Gesneriaceae (gloxinias and African violets), in the neotropical tribe Gesnerieae. Most species in this tribe are herbs or small shrubs. Their pollinators include hummingbirds, nectar bats, moths, and diurnal insects such as pollen-collecting bees and flies. A multivariate analysis of the floral morphology of these species revealed three basic floral types: diurnally opening red or orange tubular flowers, nocturnally opening light green or white bell-shaped flowers, and nocturnally opening yellow or red so-called sub-bell-shaped flowers. As you can guess, two of these floral types correspond with classic hummingbird and bat pollination syndromes and detailed pollinator observations confirm this. The third floral type is "generalized" regarding its visitors, which include bats, moths, hummingbirds, and diurnal insects. A phylogenetic analysis indicates that both bat pollination and generalized pollination have evolved twice from hummingbird pollination in the Antilles. The evolution of generalized flowers is not surprising given the low species richness of hummingbirds and nectar bats on Caribbean islands compared with mainland habitats.

A similar shift from specialized pollination to more generalized pollination has occurred in Mexican columnar cacti. In the tropics, experimental studies in Mexico and Venezuela have shown that the large, nocturnally opening white flowers of columnar cacti (Cactaceae, subfamily Cactoideae, basal eudicots) are nearly exclusively pollinated by glossophagine bats, especially the lesser long-nosed bat. In contrast, in northern Mexico and southwestern Arizona at the geographical limits of these cacti, our experimental studies have shown that the pollination systems of these cacti are more generalized. Bat pollination is less important than pollination by diurnal birds and bees in organ pipe and saguaro cacti; bat pollination is still most important in cardón cacti. Interestingly, a similar geographic trend has not been documented in paniculate agaves in Mexico and southern Arizona. Both bats and diurnal pollinators, including hummingbirds and insects, are effective pollinators throughout this area. Differences in floral biology of these two groups can perhaps explain these differences. Flowers of columnar cacti last only one night (or day in the case of saguaros), giving most diurnal pollinators a limited time in which to visit them. In contrast, flowers of paniculate agaves last for several days, giving both nocturnal and diurnal pollinators more time to visit them.

Finally, let's consider the importance of hummingbird pollination for herbaceous and shrubby montane plants in the Rocky Mountains of western North America. Only a handful of hummingbird species of the Bee clade, including Anna's, rufous, broad-tails, calliopes, and to a lesser extent black-chins, visit these flowers. Plants in this area are North American (rather than tropical) in origin and are ancestrally insect-pollinated. They occur in several plant families: three asterids (Caprifoliaceae, honeycreepers; Polemoniaceae, phlox; and Scrophulariaceae, figworts) and two basal eudicots (Caryophyllaceae, pinks; and Ranunculaceae, buttercups). About 180 of these species are hummingbird pollinated and they are very young (about 6–7 Ma old), reflecting the relatively recent arrival of Bee hummingbirds in North America. The evolution of hummingbird pollination in these plants has occurred independently at least seventy times and appears to have involved a gradual opportunistic switch between pollinators based on habitat specialization and allopatric speciation (i.e., geographic isolation) in plants. This kind of opportunism has been a common theme in the evolution of many plant-animal interactions on Earth.

The evolution of hummingbird pollination in North American plants in the genus *Penstemon* (Scrophulariaceae) has been particularly well studied. All members of this genus have tubular flowers, but those pollinated by bees have wider corollas and are shorter and blue in color compared with the narrower and longer red flowers of hummingbird-pollinated species. Overall, this genus contains about three hundred species and transitions from bee or wasp pollination to hummingbird pollination have occurred independently up to twenty times. Examples of the reverse situation, as occurs in *Ruellia* and many other tropical lineages, are unknown. In addition, once they have evolved, species of North American hummingbird-pollinated *Penstemon* have given rise to very few additional hummingbird-pollinated species, also in contrast to some tropical plant lineages (e.g., epiphytes in the pineapple family, Bromeliaceae, a monocot). Perhaps this reflects the young evolutionary age of this interaction. It also likely reflects that fact that all North American hummingbirds are migratory and can potentially carry pollen long distances between montane plant populations whereas many (but certainly not all) tropical hummingbirds are nonmigratory (see pages 83–84) and are short-distance pollen movers. Long-distance pollen movements can reduce genetic divergence among populations and rates of speciation compared with short-distance pollen movements.

WRAPPING IT ALL UP

Now that I've completed a detailed examination of the evolution of new-world hummingbirds, nectar bats, and their major food plants, it's time to conclude parts 3–5 with a few summary statements. I'll first do this with one picture that is certainly worth millions of years of evolutionary history. In figure 24 we see a summary of the evolutionary history of North American plants and their hummingbird pollinators. It includes estimates of the ages of particular species or groups of plants that are pollinated by particular species or groups of Bee hummingbirds. In viewing this figure, we should strive to recall everything that has gone on evolutionarily in these plants and hummingbirds to arrive at this current situation: how much flower evolution it has taken to create these interactions and how much time has elapsed during this process. Although I have not found a similar figure that illustrates the evolutionary history of bat-pollinated flowers and their mammalian pollinators, I'm sure that a similar kind of summary is certainly possible. A major difference between the hummingbird and nectar bat scenarios, however, is that all of the bat/plant connections involve tropical lineages in both groups. Neotropical nectar bats do not venture far enough into the United States to interact with temperate plants. The general message here, at least for North American plants and hummingbirds, is that the interactions we see today are the result of a few million years of evolution during which much has changed in the morphology, physiology, and behavior (sensu lato) of these organisms. This evolution has involved divergence from their common ancestors several millions of years ago and, in many aspects, convergence toward similar solutions to particular evolutionary challenges. The major challenges here include efficiently obtaining and processing nectar on the hummingbird side and effectively attracting these pollinators on the plant side. The end result of this evolution has been the adaptive radiation of particular groups of plants and their nectar- and pollen-seeking pollinators as a result of coevolution between them.

While ever-more-detailed information about "big picture" or macroevolutionary events such as those illustrated in figure 24 are continually forthcoming, "small picture" or microevolutionary events based on our ever-increasing knowledge about the genetics involved in particular kinds of adaptations in hummingbirds and nectar bats are also occurring. For example, a recent paper in the journal *Science* reports that the inactivation of one gene (FBP2, which codes for an enzyme that controls glucose production in muscles [you needn't worry about its tongue-twisting chemical name!]) in ancestral and other hummingbirds has resulted in increased glucose and mitochondrial production and higher levels of aerobic respiration in muscles needed for hovering. This deactivation has not been observed in a large sample of other kinds of

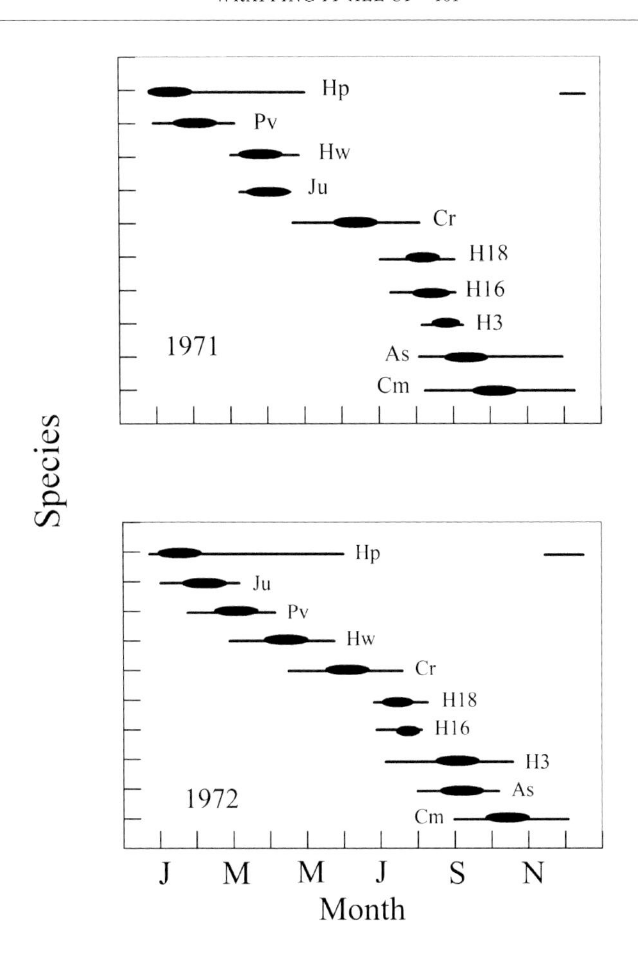

FIGURE 23 The flowering phenology of plants pollinated by hummingbirds at the La Selva research station in Costa Rica. Each horizontal line indicates the flowering phenology of a different species with major flowering periods indicated by the black ellipses. From Fleming and Kress (2013) with permission.

birds, including swifts and other Strisores, which indicates that it likely is a unique adaptation in hummingbirds.

For their size, bats, including many phyllostomids, are unusual among mammals for their exceptionally long life spans. Nectar bats, for example, are likely to live six or seven years or more whereas similar-sized rodents live a fraction of this. A recent genetic-based study of twenty-six species of bats from several families explores a possible reason for this. This study focused on the methylation of DNA, a process that involves the addition of methyl groups (i.e., CH_3) to DNA nucleotides such as cytosine without changing the gene sequence while changing its activity. When methylation occurs near genetic transcription factors that affect biological processes by turning genes on and off, a number of processes can be affected including aging, immune responses, and cancer suppression. Methylation of DNA occurs in all of us as we age. But in bats, as well as in humans, this process occurs at a rate that is significantly slower than in other mammals. And in this study, degree of DNA methylation of genes associated with aging turned out to be an accurate predictor of a bat's known age. The basic message here is that, like hummingbirds, bats differ significantly from other mammals in a number of important biological processes, including their rates of aging. Long life spans, in turn, make nectar bats especially effective dispersers of pollen and the genes they contain. Their long lives allow them to acquire intimate knowledge about their environments, including the seasonally changing locations of their food resources. As a result, they can be efficient, if not optimal, foragers.

6

PALEOTROPICAL NECTAR-FEEDING BIRDS AND BATS AND THEIR FOOD PLANTS

OVERVIEW OF DIVERSITY, DISTRIBUTION, AND EVOLUTION OF THESE VERTEBRATES

Convergence, which can be defined as similar morphological, physiological, and behavioral features in distantly related organisms, is a common theme in the evolution of life on Earth. Convergence often reflects the evolution of similar solutions to particular biotic and abiotic challenges that species face in different parts of the world. In sum, it reflects the fact that evolution of life on Earth tends to be redundant and that there may be limited ways in which organisms can evolve under a given set of environmental conditions. Familiar examples of convergence include the bipedal (hopping) rodents and spiny succulent plants in new- and old-world deserts.

Since convergent evolution happens, it should not be surprising to learn that the nectar birds, bats, and flowers that I've described in detail for the neotropics have their Paleotropical counterparts. Many old-world flowering plants also depend on nectar-seeking birds and bats for their reproductive success. But despite general similarities between the New and Old Worlds in many aspects of these interactions, they differ in many details because of historical differences in the geology, climate, and biotas of these two vast tropical and subtropical regions.

For one thing, the Paleotropics, which includes Africa, India, southern Asia, various islands in Indonesia and the South Pacific, and Australasia, is much more diverse geographically and historically than the neotropics. As a result, it's not surprising to know that more families and species of flower-visiting birds have evolved in the Paleotropics than in the neotropics (table 1). As in the neotropics, however, only one family of old-world bats (flying foxes and their relatives, Pteropodidae) contains a few morphologically specialized nectar-feeding species. Likewise, there is much hemispheric overlap in the plant families whose flowers are pollinated by birds and bats, but there are also a number of vertebrate-pollinated plant families that occur only in one hemisphere or the other. For example, endemic hummingbird-pollinated plant families in the New World include Bromeliaceae, Cactaceae, and Heliconiaceae (this ignores a few old-world species), Their old-world counterparts include Asphodelaceae (aloes) and Proteaceae (banksias). Similarly, endemic new-world bat-pollinated families include Asparagaceae (agaves), Bromeliaceae, Cactaceae, and Marcgraviaceae (neotropical climbers that are also pollinated by hummingbirds). Their old-world endemic counterparts include Asphodelaceae, Musaceae (bananas), Pandanaceae (screw pines), and mangroves in the traditional family Sonneratiaceae (now included in family Lythraceae).

Whereas hummingbirds—the major avian pollinators of neotropical plants—are nonpasserines, most, but not all, of the major Paleotropical avian pollinators are passerines. These include honeyeaters (Meliphagidae), sunbirds (Nectariniidae), flowerpeckers (Dicaeidae), and non-passerine lories and lorikeets (Psittaculidae) (table 1). Thus, when we lived in Panama and Costa Rica, flowers around us were visited during the day nearly exclusively by hummingbirds whereas when we lived in tropical Australia our common avian pollinators included several species of honeyeaters and noisy flocks of rainbow lorikeets.

Let's now look at some of the major features of these birds—their evolutionary history, diversity, and general biology—to see how similar or different they are from their neotropical counterparts. Surprisingly, along with falcons, parrots appear to be the closest nonpasserine relatives of the large avian order Passeriformes (songbirds). Several families of parrots are now recognized and lories and lorikeets are advanced members of the old-world family Psittaculidae, which evolved in Australasia beginning about 50 Ma ago. The clade containing lories and lorikeets is about 10 Ma old and likely first evolved in New Guinea. It contains nineteen genera and about 61 species (table 1). At 50 Ma, honeyeaters (Meliphagidae) are among the oldest members of the Passeriformes. This family is very diverse and contains fifty-three genera and about 189 species (table 1). Like lories and lorikeets, it also evolved initially in

Australasia, which is still its center of highest species richness. Sunbirds (Nectariniidae) and flowerpeckers (Dicaeidae) are sister families and are advanced members of the Passeriformes. They contain sixteen genera and 145 species and two genera and 50 species, respectively. Their ages are about 35 Ma. Both families likely first evolved in Southeast Asia, but the major radiation of sunbirds has occurred in Africa where about two-thirds of its species live; greatest species richness of flowerpeckers occurs in New Guinea.

A group of 14 species of bats classified in six genera of pteropodid bats are the old-world nectar-feeding counterparts of new-world phyllostomid nectar bats (table 1). However, unlike the phyllostomids, which are clustered within two subfamilies, the pteropodid nectar feeders are scattered throughout this family and occur in four subfamilies. Family Pteropodidae evolved in Southeast Asia about 25 Ma and most species of its nectar bats still inhabit Southeast Asia and Australasia. Only one species lives in Africa. The ages of pteropodid nectar bats range from 22 to 6 Ma with the most common species (dawn, long-tongued, and blossom bats) being about 16–17 Ma old.

AN OVERVIEW OF THEIR MAJOR BIOLOGICAL FEATURES

Major features of these birds and bats are summarized in tables 6 and 7. Compared with their new-world counterparts, these nectar feeders tend to be larger, and many (most?) of them do not routinely hover at flowers to feed. Of these birds, sunbirds are most likely to hover at flowers, especially at those that do not provide a landing place for them. Since old-world sunbirds and honeyeaters are more similar to new-world hummingbirds than flowerpeckers and lories. I will concentrate on these two families here. I have had little personal experience with sunbirds but have watched a few species visiting flowers in Africa, Malaysia, and Australia. The year we lived in Australia, in contrast, I noted having seen thirteen species of honeyeaters during travels and fieldwork in tropical Queensland. These ranged from the small red-headed honeyeater (about 11 g) to the large noisy friarbird (about 136 g).

SUNBIRDS

Hunter's sunbird from eastern Africa can introduce us to this family (figure 25A). I watched it feed while perched at aloe flowers in Tarangire National

TABLE 6 Summary of some major biological features of old-world nectar-feeding birds

FEATURE	LORIES AND LORIKEETS	HONEYEATERS	SUNBIRDS	FLOWERPECKERS
Range of sizes (g)	20–240	7–244	4–38	5–13
Bills and tongues	Strong, curved bills like other parrots; tongues are brush tipped	Bills slightly decurved; tongues brush tipped	Bills long, thin, and decurved; tongues are tubular and brush tipped	Bills not elongated; tongues not brush tipped
Plumage characteristics	Brightly colored	Noniridescent but many have bright-colored skin or feathers on head and upper breast	Iridescent and brilliantly colored	Noniridescent; mostly browns or grays but some have red or yellow plumage
Social behavior	Monogamous but highly gregarious; flock foragers	Larger species are strongly territorial	Many are territorial but spiderhunters are trapliners	Flock foragers outside the breeding season
Diet	Fruits and seeds in addition to much nectar	Broad diets but many are strong nectar feeders	Primarily insects but many are avid nectar feeders	Mainly fruit (mistletoe berries) but also nectar and insects
Movements	Highly mobile outside the breeding season	Can be highly nomadic	Often nomadic	Nonmigratory?

TABLE 7 Summary of some major biological features of old-world nectar-feeding bats

FEATURE	PTEROPODID NECTAR BATS
Range of body sizes (g)	23–82
Rostrum and tongues	Rostrum somewhat elongated, teeth reduced in size and number; tongues protrusible and tipped with a brush of "hairs"
Social behavior	Many are strongly territorial and solitary foliage roosters; dawn bats are gregarious and roost in caves
Movements	Not likely to be migratory but opportunistic nectar feeders can be nomadic

Note: Many nonspecialized pteropodids (e.g., species of *Pteropus*, *Rousettus*, and *Cynopterus*) are opportunistic and common nectar feeders. They are usually much larger than the specialized pteropodid nectar bats.

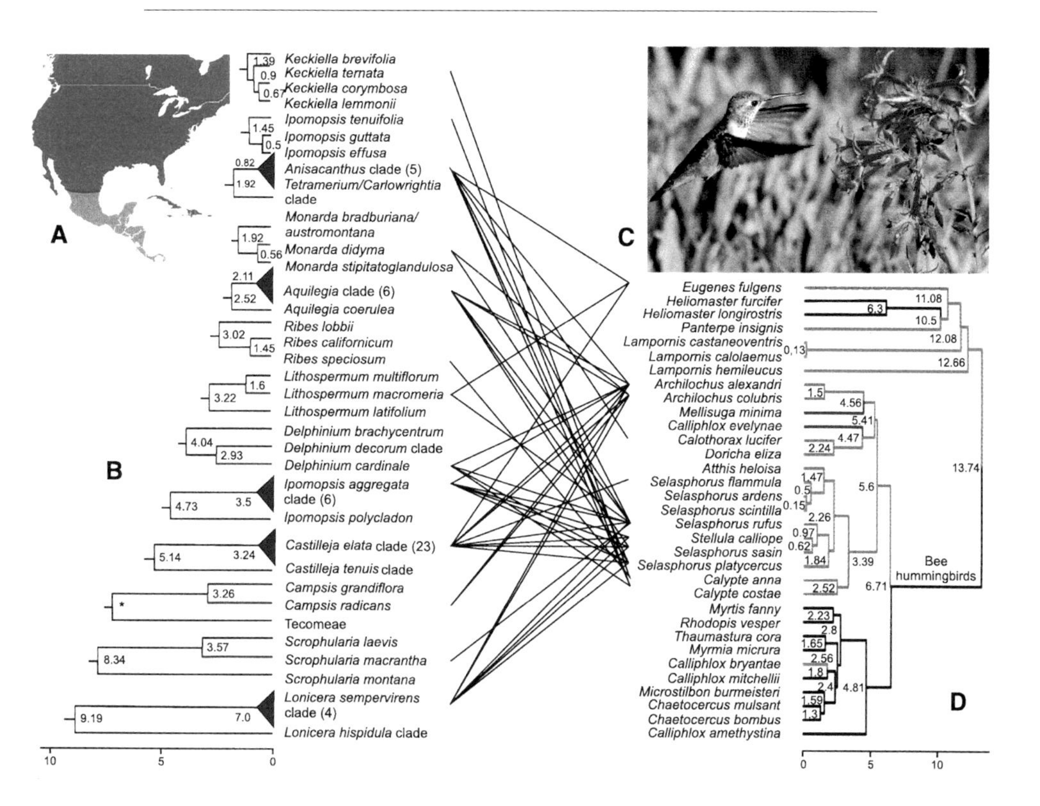

FIGURE 24 An evolutionary tanglegram indicating feeding connections between a series of hummingbirds and their North American food plants. The horizontal scales at the bottom of the figure are age in millions of years. From Abrahamczyk and Renner (2015) with permission.

Park in eastern Tanzania. Sexes are dimorphic in plumage in this relatively small (about 12 g) bird. Males are mostly black with an iridescent green crown and "mustache" and a red chest. Females are mostly brown with a brown-streaked breast. It is a bird of open woodlands and semiarid grasslands where it feeds on a mixture of insects, fruit, and nectar. Flower sources include acacias (Fabaceae), aloes, and various trees (e.g., other members of Fabaceae and Bignoniaceae [e.g., the sausage tree]). Its seasonal movements are poorly known but likely include short latitudinal movements after the breeding season.

Sunbirds are among the most colorful old-world birds and are the closest ecological analogues of hummingbirds. They are slender, lightly built birds, and most species have long, thin, and slightly decurved bills. All of them have long extensible tongues whose tips are split and brush covered. Though small, on average they weigh twice as much as hummingbirds, and because they are perching (rather than hovering) birds, they have long, strong legs. Like hummingbirds, many sunbirds are sexually dimorphic, and males of most species have brilliant, iridescent plumages featuring blue, purple, or green heads and backs as well as bright reds and yellows on their backs and bellies, respectively. As in hummingbirds, this iridescence is structural in nature as a result of the interaction between sunlight and keratin and melanin granules in their feathers.

Sunbirds also share some behavioral traits with hummingbirds. During the breeding season, for example, pairs of birds (rather than single individuals) are territorial and defend patches of flowers against intruders. After the breeding season, many sunbirds, especially those living in dry habitats, wander widely in search of flowers. A group of Asian sunbirds known as spiderhunters (e.g., birds in the genus *Arachnothera*) resemble hermit hummingbirds and are trapliners rather than territorialists. Like hermits, they visit flowers in forest understories and are larger and duller colored than most other sunbirds.

Unlike hummingbirds, sunbirds are typical passerines with monogamous breeding systems and biparental care of the nestlings. Most species place their grassy globular nests in low shrubs or trees, but, like hermit hummingbirds, spiderhunters attach their nests with spiderweb material to the undersides of palm or banana leaves.

The diets of sunbirds include many insects as well as nectar and pollen. Unlike hummingbirds, however, they have the ability to extract amino acids from pollen grains in their stomachs to supplement the nitrogen in their diets. But, as an adaptation to the chronically low nitrogen levels in their diets, both of these families of birds have much lower daily nitrogen requirements than other similar-sized birds. Also, in contrast to hummingbirds, sunbirds can adjust the amount of water their intestines absorb to reduce water stress on their kidneys. Finally, because they don't usually hover while visiting flowers, sunbirds have substantially lower daily costs of living than similar-sized hummingbirds. But, like nearly all passerines, they do not undergo torpor at night to reduce their overall daily energy costs.

Overall, sunbirds display an impressive number of similarities with hummingbirds in terms of their morphology, behavior, and physiology. Especially striking are the similarities between hermit hummingbirds and spiderhunters. These similarities are excellent examples of convergent evolution between distantly related animals. Nonetheless, these birds also differ in a substantial number of traits that reflect their different evolutionary histories. Sunbirds, after all, are passerine songbirds whereas hummingbirds are nonpasserines. Being passerines, they possess traits such as monogamous breeding systems, the inability to hover for substantial periods, and strong legs for perching and walking on the ground that are absent in hummingbirds. Also, forest-dwelling species are known to forage in mixed-species flocks, behavior that is unknown in hummingbirds.

HONEYEATERS

In contrast to sunbirds, honeyeaters do not closely resemble hummingbirds (or sunbirds), at least in many aspects of their external morphology. For starters, they are up to seven times heavier than hummers and do not hover when feeding at flowers (table 6). Instead, they perch on or near flowers to feed. In addition, most species are dull colored (green, gray, or dark brown with black or white markings), but some have bright red, yellow, or blue faces. Denoting their nectar-based diets, most honeyeaters have relatively long, thin, and slightly down-curved bills and long tongues with brushy tips.

The life history of Lewin's honeyeater, which is a common species that lives along Australia's east coast, can introduce us to these birds. It was the species that I saw most often when we lived for a year in Townsville, northeast Queensland. This species is about the size of an American robin (up to about 50 g), and both sexes have similar plumages. It is mostly gray with olive-colored wings and inner tail feathers (figure 25B). A bright white stripe runs from its eye to its slightly curved bill, and it has a large crescent-shaped yellow spot behind its eye. This honeyeater is mainly a forest dweller but is also common in urban areas that contain flowering and fruiting plants. Its diet is mostly fruit, but it also eats many spiders and insects and visits flowers for nectar. It is a relatively sedentary species, but like many other plant-visiting birds, it sometimes undertakes short movements among habitats seasonally in search of rich food patches.

Reflecting their passerine status, most honeyeaters have monogamous breeding systems with biparental care (table 6). Many species are highly territorial during the breeding season, and size-based dominance hierarchies sometimes exist in habitats where several species co-occur. Large species tend to dominate the richest patches of flowers while small species feed in patches of lower flower density. During the nonbreeding season, species living in arid habitats are often highly nomadic and move substantial distances while tracking changes in the locations of their favorite food species. Species living in wetter habitats, in contrast, tend to be more sedentary year-round.

Some of the digestive and metabolic adaptations of honeyeaters are similar to those of hummingbirds and sunbirds. For example, they all have lower mass-specific nitrogen requirements than expected for their size. Like sunbirds, however, honeyeaters are better able to extract the contents of pollen grains, including their amino acids, than hummingbirds. Like sunbirds, they can adjust the amount of water their intestines absorb to reduce water stress on their kidneys. Also like sunbirds, honeyeaters do not use torpor to reduce their daily energy costs. Nonetheless, because they don't hover, their energy costs are about 43 percent lower than those of similar-sized hummingbirds.

Overall, honeyeaters are among the most abundant and conspicuous birds throughout much of Australia. Their evolutionary success can be attributed in large part to their generalized feeding habits. While they clearly possess adaptations for nectar feeding, which is the ancestral feeding condition in the family, the diets of most species are broad and include many fruits and insects. Fruit eating is especially common in larger species whereas nectar feeding is especially common in smaller species. Compared with hummingbirds and sunbirds, honeyeaters live in a part of the world (Australasia) where plant-based food resources such as fruit and flowers exhibit an especially high degree of spatial and temporal variability compared with the neotropics and Africa. Boom and bust years are common in their food resources as a result of highly variable annual rainfall. Thus, unlike many other nectar feeders, feast or famine are challenges that these birds face on a regular basis. As a result, natural selection has placed a premium on feeding flexibility. Part of this flexibility involves nomadic lifestyles, a life history feature that is also common in Australian flower-visiting and fruit-eating pteropodid bats.

FLOWERPECKERS AND LORIKEETS

Flowerpeckers and nectar-feeding lories and lorikeets are two additional families of old-world flower-visiting birds that I'll touch on briefly here. Flowerpeckers belong to a family (Dicaeidae) that is sister to the sunbirds. Other than their evolutionary relationship, however, flowerpeckers and sunbirds have little else in common. Whereas sunbirds are slender, brilliantly colored birds with long curved bills, flowerpeckers are basically small, dull-colored, dumpy birds with short, pointed bills. Although they occasionally visit flowers to feed on nectar, they are better known for their fruit-eating habits and are sometimes called mistletoe birds. The most important fruits in their diets are mistletoe berries, and they can be considered to be specialists on these very sticky fruits. Because they move quickly from one fruiting plant to another, they have been described as being very energetic and almost frenetic foragers. In tall forests they feed mostly in the canopy but feed down lower in shorter, more open habitats. Like many other Australasian birds, they form noisy flocks and forage widely during the nonbreeding season.

Lories and lorikeets are members of subfamily Loriinae of the old-world parrot family Psittaculidae. Whereas most parrots are seed and fruit eaters, lorikeets are dedicated flower visitors that feed on nectar and pollen. They are rather slender bodied and have a slender bill that houses a brush-tipped tongue, like most other nectar-feeding birds and unlike other parrots.

The rainbow lorikeet (figure 25C) is perhaps the best-known member of this clade. It is the most common nectar-feeding parrot along the eastern coast of Australia and was a common seasonal visitor to our yard in Townsville. This relatively large bird (up to 169 g) is unmistakable because of its colorful plumage that features a dark blue head and belly, bright orange upper chest and underwings, and a green back and tail. Males and females look alike. These birds inhabit a variety of lowland forests where they feed heavily on the nectar and pollen of many kinds of flowers, especially those in the eucalypt (Myrtaceae) and banksia (Proteaceae) families. Like other lorikeets, the rainbow lorikeet is very noisy both whenever it flies and while it is feeding. It screeches when flying and constantly chatters when feeding. It is monogamous, and pairs usually place their nest in tree cavities or holes. Like other species in this subfamily, this lorikeet is nomadic and travels in flocks in search of patches of flowering plants after the breeding season.

Studies of the digestive physiology of rainbow lorikeets indicate that, like other nectar-feeding birds, the transit time of watery nectar through their gut is quite rapid (about eighty-eight minutes) and that passive absorption (rather than active transport via enzymes) of sugars such as glucose in the small intestine is very high (at least 90 percent). The digestion of the pollen they ingest in their stomachs, however, is very low (about 7 percent). So they need to supplement their nectar diet with other kinds of food (seeds, leaves, and fruit) to obtain the protein and other nutrients they need. This once again emphasizes the nutritional tradeoffs that nectar feeders face. On the one hand, nectar is often readily available, is well advertised, and is an excellent source of energy. But it's nutritionally incomplete, which forces most nectarivores to use other food sources for a complete diet. By providing an easily obtained but nutritionally incomplete food reward, plants likely benefit by increasing the mobility of their pollen and the genes it contains.

NECTAR BATS

Plant-visiting bats in the old-world family Pteropodidae are the ecological counterparts of nectar- and fruit-eating members of the new-world family Phyllostomidae. Although the diversity of morphologically specialized pteropodid nectar bats is modest (table 1), this doesn't really reflect how many species of pteropodids actually visit and pollinate flowers. Many morphologically nonspecialized pteropodids (e.g., many species in the genus *Pteropus* as well as those in *Cynopterus* and *Rousettus*) are also frequent flower visitors. For example, when we lived in Queensland, Australia, it was common to see and hear black flying foxes squabbling among themselves in the canopies of flowering eucalyptus trees. Similarly, when my son, a group of zoology undergraduate students, and I "staked out" a flowering blue gum eucalyptus tree to monitor the comings and goings of diurnal and nocturnal flower visitors, our daytime visitors included three species of lorikeets, several species of honeyeaters, and a sunbird. As soon as the sun set, little red flying foxes began to feed in the tree, which turned out to be a 24/7 feeding station. Other blue gum trees in the area were also being visited by many nectar feeders.

My personal contact with specialized pteropodid nectar bats has been very limited, in part because I was mainly interested in fruit-eating species when we lived in Australia. The only nectar eater that I captured in mist nets was the Queensland blossom bat. In October–December 1987 I caught several adult males and females weighing 16–20 g in creek beds and around fruiting trees in north Queensland. Every individual was extremely feisty and very prone to bite my gloved hand. Because of their seemingly crazed behavior, my colleague Hugh Spencer has called these bats "Psychonycteris" (rather than their real name *Syconycteris*). One other nectar bat, the Northern blossom bat, a close relative of the Queensland blossom bat, lives in northern Australia, but I didn't catch any during my fieldwork. Both species also live in New Guinea.

During a stay at the Xishuangbanna Tropical Botanical Garden in Yunnan Province, China, I watched another pteropodid nectar bat, the cave-dwelling dawn bat, visit flowers of the tree *Markhamia stipulata* (Bignoniaceae, an asterid) on two nights (figure 25D). This bat is much larger than Australian blossom bats and weighs up to 82 g. *Markhamia* flowers are large and broadly tubular; they are placed at the tips of long stalks extending vertically away from the foliage. When I watched several flowers on one tall tree, they were being visited by solitary bats that contacted flowers for about one second before flying away. During these visits, each bat landed on a flower and did not hover. Single brief flower visits seem to typify both old-world and new-world nectar bats.

Although they are not closely related, all pteropodid nectar bats have a slightly elongated and slender snout, a long slender tongue with a brushy tip, and small teeth—features that are also found in phyllostomid nectar bats. Reflecting their higher species richness, however, new-world nectar bats exhibit a much greater array of skull shapes than their old-world counterparts. Nothing like the very long-snouted banana bat or the extremely long-tongued tube-lipped nectar bat exists in the Old World.

FIGURE 25 Four old-world nectar feeders: A. Hunter's sunbird (*Chalcomitra hunteri*), B. Lewin's honeyeater (*Meliphaga lewinii*), C. rainbow lorikeet (*Trichoglossus haematodus*), and D. dawn bat (*Eonycteris spelaea*). A is from Africa, B and C are from Australia, and D is from SE Asia. Photo credits: Ted Fleming (A), Adobe Stock (B, C), and Merlin Tuttle (D).

Pteropodid bats differ from all other bats in being non-echolocators. As a result they rely solely on vision and smell to locate flowers and fruit. Bat biologists have long debated whether the absence of echolocation is the ancestral condition in all bats and whether it has been lost in pteropodids. Because the ability to produce and detect high-frequency sounds occurs in some shrews (close relatives of bats) and most bats, it is likely that the ability to echolocate is ancestral in bats and that it has been lost secondarily in pteropodid bats, which rely on vision rather than echolocation to find their food. If this is true, then the evolutionary history of bats has followed two different sensory trajectories. Bats that actively pursue prey during flight, which is likely to have occurred in the earliest fully flying bats, would have used echolocation to find and capture prey; most of these kinds have small eyes. In contrast, bats that never pursued prey during flight but that instead fed on fruit and flowers would have experienced selection for enhanced vision and large eyes at the expense of the ability to echolocate. Plant-based diets are thus probably ancestral in pteropodids whereas they are a derived condition in phyllostomids.

Another striking difference between old- and new-world nectar bats is their roosting behavior. Whereas most phyllostomids are gregarious and live in colonies of a few dozen to tens of thousands in caves, mines, culverts, and hollow trees, most specialized nectar-feeding pteropodids roost either solitarily or in small groups in the foliage of forest canopies. The dawn bat is an exception to this. It lives in colonies containing hundreds to thousands of individuals in caves, and bats travel substantial distances (i.e., tens of kilometers) to their feeding areas. In contrast, foliage roosters such as the Queensland blossom bat frequently change roost sites and commute only a kilometer or so from roosts to their feeding areas. This and other species of foliage-roosting nectar bats tend to forage in small feeding areas (e.g., only about 5 ha), and it is likely that adult males are territorial and defend these areas from other conspecifics. As in many territorial hummingbirds, females and juveniles of territorial nectar bats have to sneak into rich feeding patches when males are chasing other bats. As a result, they are likely to be more important pollen dispersers than territorial males. A similar situation occurs in the much larger opportunistic nectar-feeding species of *Pteropus* (flying foxes), which often live in large colonies in tree canopies. Weighing several hundred grams to a kilogram or more, some of these bats (probably adult males) establish small feeding territories in the canopies of flowering eucalyptus trees. Juveniles and females are forced to "sneak" into these territories to obtain nectar and pollen. And, like their blossom bat counterparts, they are the ones that move pollen from one plant to another. Australian bat biologists Les Hall and Greg Richards have called this a "residents and raiders" pollen dispersal system, which also occurs when these bats are feeding on fruit produced by canopy trees.

Physiological studies of pteropodid nectar bats are far fewer than those for hummingbirds and phyllostomid nectar bats, but available information suggests that all of these animals, including much larger Egyptian fruit bats weighing 150 g, quickly metabolize the sugars they ingest and use this energy to directly fuel their flight muscles. In addition, studies based on carbon stable isotope analysis (see page 42) suggest that rather than obtaining the nitrogen they need from eating insects, as some hummingbirds and phyllostomids do, pteropodids are capable of meeting their daily nitrogen needs, which tend to be lower than expected based on their size, by consuming nectar, pollen, and fruits.

In contrast to phyllostomid nectar bats, torpor occurs in at least two species of pteropodid nectar bats. In the northern (or Queensland) blossom bat in Papua New Guinea, for example, females and small males living in lowland forests undergo torpor when air temperatures fall below 30°C whereas large males and all individuals living in the mountains do not. Similarly, the common blossom bat also undergoes torpor in the lowlands of New Guinea but not in the highlands when air temperatures drop below 30°C. When thermoregulating, both of these species have body temperatures of 35°C–36°C, several degrees lower than most birds. While in torpor their body temperatures drop to about 4°C above ambient temperatures. Their basal metabolic rates are also lower than those of hummingbirds and terrestrial mammals of similar size. Torpor in lowland females has been suggested to be an adaptation for prolonging the life of sperm that they store in their reproductive tracts prior to fertilization of their eggs.

This brief survey of old-world nectar bats indicates that they share many morphological and physiological traits with their new-world counterparts. These similarities again illustrate the phenomenon of convergent evolution. Behavioral differences between these two groups, however, support the idea that this convergence does not apply to all aspects of their lives. For example, nectar-feeding pteropodids are generally much more intolerant of and aggressive toward conspecifics and thus more closely resemble the social behavior of hummingbirds than that of phyllostomid nectar bats. A further striking difference is the lower taxonomic diversity and its correlated lower morphological diversity in pteropodid nectar bats compared with their new-world counterparts. Both groups of bats are similar in evolutionary age (i.e., 20–25 Ma), but phyllostomids have undergone a much greater diversification of morphologies and species. A similar hemispheric difference exists in the diversification of avian nectar feeders, which suggests that ecological opportunities for diversification have historically been greater in the New World than in the Old World for these kinds of animals. In a number of my publications, I have suggested that hemispheric differences in the spatial (geographic) and temporal (year-year) predictability of food sources have been important factors behind these differences.

How does the spatiotemporal predictability (STP) of food resources affect the evolution of morphological and taxonomic diversity in these (and many other) animals? The short answer is that high levels of STP likely favor the evolution of feeding specialization whereas low levels of STP favor the evolution of feeding generalization. When food resources are highly reliable in space and time, then consumers can afford to specialize on them, and by definition, they will likely occupy rather narrow ecological food niches. When food resources are less reliable in space and time, consumers must have broader food diets and niches in order to survive. Intuition, as well as many theoretical studies, suggest that more specialized species can co-occur in communities than generalized species, a pattern that we see in hemispheric comparisons of communities of nectar-feeding (and fruit-eating!) birds and bats. In addition, over time, the evolution of specialized species will generate more taxonomic diversity than the evolution of generalized species. In sum, it appears that hemispheric differences that we see in many of the patterns of morphological, ecological, and taxonomic diversity of nectar-feeding birds and bats likely reflect historical differences in the STP of their food resources. New-world species appear to have evolved in a world of greater resource reliability than old-world species. Providing additional support for this hypothesis is that we can see the same kinds of hemispheric differences in fruit-eating birds and mammals. Together, these broad patterns reflect the obvious fact that we live on a planet characterized by significant geographic variation in conditions that affect the evolution of life. This variation clearly exists in our world today (e.g., compare the biotas of tropical rainforests with those of deserts), and it has undoubtedly existed throughout much of Earth's history.

THE KINDS OF FLOWERS POLLINATED BY OLD-WORLD BIRDS AND BATS

BIRD FLOWERS

How similar are the food choices of new- and old-world nectar-feeding birds and bats? Do they feed at flowers produced by the same plant families, and do they visit flowers of similar morphology (figures 20 and 21)? First let's compare hummingbirds with sunbirds. Hummingbirds are known to visit flowers in at least ninety-five plant families. As we've already seen, the top five families and their growth forms are: Fabaceae (especially mimosoid [mimosa-like] shrubs and trees), Acanthaceae (herbs), Bromeliaceae (epiphytes), Rubiaceae (shrubs), and Lamiaceae (herbs). Whereas four of these families produce tubular flowers, mimosoid legumes produce shaving-brush flowers characterized by an abundance of stamens and stigmas. In terms of plant growth habit, hummingbirds are most likely to visit flowers of herbs and epiphytes and are least likely to visit those of trees. And the overwhelming majority of the flowers they visit are tubular in shape. Sunbirds are also known to visit flowers in at least ninety-four families. Their most "popular" families and their growth forms include Fabaceae (mimosoid shrubs and trees), Rubiaceae (shrubs), Lamiaceae (herbs), Loranthaceae (hemiparasitic mistletoes), and Bignoniaceae (trees and vines). They also heavily visit aloes (Asphodelaceae). While sunbirds visit many flowers produced by herbs and shrubs, forest-dwelling species also visit many flowers produced by trees. Like hummingbird flowers, most of the flowers visited by sunbirds are tubular in shape. In sum, both families of birds visit many of the same kinds of flowers produced by many of the same families. Nonetheless, not surprisingly because they occur in different hemispheres, sunbird diets also include plants from families not visited by hummingbirds. These include Araliaceae (ivies and ginseng), Arecaceae (palms), Chrysobalanaceae (cocoa plums), Ebenaceae (persimmons), and Theaceae (camellias and tea). Although all of these families have pantropical (worldwide tropical) distributions, they are only pollinated by birds in the Old World. Available data also suggest that most species of sunbirds have broad diets and that they are less likely to have evolved specialized relationships with particular kinds of flowers than hummingbirds such as hermits which are *Heliconia* specialists and swordbills which are *Passiflora* (passion vines) specialists.

Honeyeaters and lorikeets appear to be more generalized flower visitors than either hummingbirds or sunbirds. Honeyeaters, for instance, are known to visit flowers in only forty families of plants. Both honeyeaters and lorikeets are more likely to visit flowers produced by trees and shrubs than sunbirds and most of these flowers are nontubular in shape. Because they are open and cup shaped or shaving brushes with lots of stamens, they are accessible to many kinds of birds in addition to more specialized species. Important plant families for lorikeets include eucalypts (Myrtaceae, trees) whereas those of honeyeaters include eucalypts, banksias (Proteaceae, trees and shrubs), citrus (Rutaceae, trees and shrubs), and Fabaceae (trees and shrubs). Because eucalypts are the dominant trees in many Australian forests, lorikeets and honeyeaters (and nectar bats) often have vast populations of flowers to visit—a situation that hummingbirds and sunbirds never encounter.

Finally, because most old-world nectar-feeding birds (and all bats) do not hover, they need to perch on or near flowers to feed. As a result, many of their flowers have evolved strong structures such as substantial branches or floral bracts (as in Asian vertebrate-pollinated banana flowers) or modified inflorescences as in the terrestrial iris *Babiana* in South Africa to provide greater access to their flowers.

BAT FLOWERS

Our review of flowers visited by phyllostomid and pteropodid bats indicates that these occur in sixty-seven families of which twenty-six are exclusively visited by phyllostomids and twenty-three are exclusively visited by pteropodids; eighteen families are visited by both families of bats. Thus, bats worldwide visit flowers in fewer families than hummingbirds and sunbirds. Except for basal angiosperms, these families are scattered throughout angiosperm phylogeny (figure 1). The most "popular" plant families containing flowers pollinated by pteropodids are Myrtaceae (eucalypt trees), Bignoniaceae (trees and vines), Arecaceae (palms), Malvaceae, subfamily Bombacoideae (trees), and Fabaceae (trees). Thus, unlike phyllostomid nectar bats, most of the flowers visited by pteropodids are produced by trees rather than by vines or epiphytes. In addition, most phyllostomid flowers are tubular or cup shaped whereas those visited by pteropodids are either cup shaped or shaving brushes. Overall, because they limit access to their nectar, the tubular flowers visited by phyllostomid nectar bats tend to be more specialized morphologically than those visited by pteropodids.

A study of the diet of the cave-dwelling dawn bat in Thailand gives us a good introduction to food choices made by a common pteropodid nectar bat. Over the course of a year, pollen found in fecal and fur samples revealed that these relatively large bats fed on at least eleven flower species from seven families of plants. Except for the megaherb *Musa* (wild banana), all of these species were trees. On any given night, individuals fed on up to six different species of flowers, which suggests that they can be very opportunistic feeders and pollinators. Flowers from two species of *Parkia* (canopy trees) and *Musa* were heavily visited throughout the year. These species plus several others (e.g., durian and kapok) are economically very important species as sources of food or fiber for humans in Asian forests.

Another study conducted in southern Thailand extends our knowledge about diet choices of Asian nectar bats. This study focused on three morphologically specialized species (the dawn bat and two species of small blossom bats) and four species of opportunistic flower visitors (three species of short-nosed fruit bats and Leschenault's rousette, another fruit bat). Of the three specialized species, the dawn bat was a feeding generalist, as indicated above, whereas the blossom bats were feeding specialists and concentrated their visits nearly exclusively on flowers of banana plants. Pollen was rarely found on the four opportunistic flower visitors, suggesting that nectar and pollen were not significant components of their diets.

Plant flowering phenology turns out to have an important effect on food choices made by the three specialized nectar feeders. Two major flowering patterns can be recognized in the plants visited by these species: "big bang" and "steady state." Big bang species such as durian and kapok trees produce large flower crops in short flowering seasons whereas steady state species such as bananas produce a few flowers per plant per night throughout most of the year. Furthermore, flowers of big bang species produce copious amounts of nectar compared with the much more stingy flowers of steady state species. The diet of dawn bats, a fast-flying and wide-ranging species, contains both big bang and steady state species, indicating that they feed on species exhibiting a variety of spatiotemporal patterns. In contrast, the smaller, much less wide-ranging blossom bats concentrate heavily on spatiotemporally predictable banana flowers.

GENETIC CONSEQUENCES OF PALEOTROPICAL VERTEBRATE POLLINATION

Based on my review of the effect of hummingbirds and phyllostomid bats on the genetic structure of and speciation in their food plants, we should expect to find similar effects from old-world nectar-feeding birds and bats. That is, the sociality and foraging behavior of various kinds of birds and nectar bats is likely to have genetic consequences for their food plants. Territorialists such as sunbirds, honeyeaters, and blossom bats are less likely to provide extensive pollen movement for their food plants than wide-ranging species such as lorikeets and dawn bats. Consequently, we might expect species such as bananas whose flowers are visited by territorial blossom bats to exhibit greater between-population genetic differences than those visited by more mobile species such as dawn bats. But because highly mobile dawn bats are also frequent visitors to banana flowers, they might serve to reduce genetic differences between banana populations. Opportunistic flower visitors such as flying foxes in Australia are known to fly substantial distances (many kilometers) between patches of flowering eucalyptus trees, and hence, like dawn bats in Asia, they are also likely to be long-distance pollen movers.

To start this discussion, let's look at a comparison of the effectiveness of three species of nectar-feeding vertebrates (two honeyeaters and the common blossom bat) as pollinators of the bumpy satinash tree (Myrtaceae) on the Atherton Plateau in northeastern Queensland, Australia. This large tree with its delightful name is a common member of intact and fragmented forests in this area. I was well aware of it during my Australian wanderings because of the way it presents its flowers and fruits to vertebrates. They occurred in thick clusters on its trunk rather than at the tips of its branches. This method of presentation is known as *cauliflory,* and it appears to be more common in vertebrate-pollinated and -dispersed trees in the Old World than in the New World. Why this is is currently unknown.

Bumpy satinash flowers have white petals with a shaving brush array of stamens, making them accessible to many kinds of pollinators, including insects as well as vertebrates. In one study, the two honeyeaters were more frequent flower visitors and visited more flowers per tree than blossom bats. But because bats have a dense coat of fur, they acquired more pollen per visit than birds. But most importantly for pollen movement, bats were more mobile than birds and sometimes flew up to about six kilometers between forest fragments to feed. Hence, they were substantially better pollen dispersers than birds. Consequently, we would expect that the population genetic structure of this tree is more likely to be determined by foraging movements of this bat than by honeyeaters. Genetic studies appear to support this expectation. Also, as has been reported for new-world nectar bats, because they often visit flowers in both intact forest and forest fragments, these bats are important for maintaining genetic connections among plants in human-disturbed landscapes. I will return to this topic in the Conservation section below.

I will continue this discussion of the genetic structure of bumpy satinash and other species of *Syzygium* because it highlights the roles of both pollination, the main focus of this book, and seed dispersal as mediated by fruit-eating vertebrates in determining the population genetic structure of plants. Because pollen, which is haploid (one set of chromosomes and genes), carries only one set of a plant's genes whereas seeds, which are diploid, carry two sets, it might be expected that seed dispersal, rather than pollination, has the greater effect on this structure. Many studies, however, have shown that both pollination and seed dispersal can affect a plant's genetic structure. Nonetheless, it may be that the genetic structure of trees such as various species of *Syzygium*, whose relatively large seeds are dispersed by fruit-eating pteropodid bats, is determined primarily by seed dispersal rather than by pollen movement.

Visits to *Syzygium* flowers by nectar-feeding birds and bats and the subsequent dispersal of its seeds by frugivorous bats are commonplace natural history events. But what are the long-term effects of these interactions? How have these activities affected the geographic distribution and evolution of these plants over millions of years? With the aid of modern molecular genetic tools, it is now possible to trace in detail the evolutionary history of these plants. Although we don't really know in detail the role that vertebrate pollinators and seed dispersers have played in this history, from what we know about its current natural history, it seems reasonable to make some inferences about this.

About 1,200 species of trees and shrubs have been classified in the genus *Syzygium*, making it the most diverse group of trees in the world. Although some species of this genus can now be found throughout the world's tropics because people like to eat their sweet fleshy fruits, they occur naturally from Africa through Australasia with greatest species richness in Malaysia and Australia. They are members of the myrtle family (Myrtaceae, a rosid), a family that includes eucalypts, another very diverse group of Australasian trees and shrubs that produce nonfleshy capsular fruits and shaving brush flowers that are frequently visited by many species of nectar-feeding birds and bats (see above).

By documenting the DNA nucleotide sequences of the entire genomes of many species of *Syzygium*, botanists have been able to reconstruct major features of their evolutionary history. Results of this research indicate that the ancestral home of *Syzygium* is Australia-New Guinea (where its current diversity is high). From there it underwent multiple migrations into the Indo-Pacific as far east as Hawai'i and westward into India and Africa. An evolutionary hypothesis (i.e., a phylogeny) of species in this genus suggests that the eastern colonizations predated the western colonizations. It should be noted that this entire region (except Hawai'i) is the domain of many species of fruit-eating (and flower-visiting) pteropodid bats, and it seems reasonable to infer that through their seed-dispersing behavior, these bats have played a significant role in the geographic expansion of *Syzygium* from its ancestral home. For example, island-hopping in the Indo-Pacific as mediated by seed-dispersing fruit bats (and fruit-eating pigeons) is certainly a potential mechanism for this. But we also need to remember that fruit and seed production would not have been possible without successful pollination of *Syzygium* flowers by these bats in the first place. Thus, both kinds of mutualisms (vertebrate pollination and seed dispersal) have probably played important roles in the evolutionary success of these plants. Additional support for this scenario comes from the fact that both the plant and pteropodid bat radiations have occurred within the last twenty million years, making them available to interact together for a relatively long period of time.

Bananas (*Musa*, Musaceae) are also prominent members of the understories of old-world tropical forests. They are classified in three genera containing about ninety species. As we've seen, blossom bats and dawn bats are important pollinators of some species, and their seeds are dispersed by terrestrial rodents. Plantain is a widespread bat-pollinated species whose population genetics have been studied in tropical China. Results of this study indicate that mainland populations contain high levels of genetic diversity and that they display little genetic differentiation. A comparison of genetic structure based on chloroplast (i.e., pollen) and nuclear (i.e., seed) DNA also indicated that gene flow via pollen movement was nearly four times higher than that via seed dispersal. These results indicate that nectar-feeding bats are critical for maintaining genetic connections between populations of this and probably other species of *Musa*.

Let's now examine the genetic consequences, and, ultimately, the production of new species as a result of pollination by other old-world nectar feeders, beginning with African sunbirds. Compared with hummingbirds, sunbirds are generally considered to be generalist feeders whose diets include fruit and insects as well as nectar. Results of a comparison of feeding relationships within communities of hummingbirds and sunbirds, for example, indicates that, on average, hummingbirds have more exclusive (specialized) relationships with their food plants than sunbirds. In addition, the morphological match between sunbird bills and their flowers is lower than it is in hummingbirds, again supporting the idea that sunbirds are less specialized nectar feeders than hummingbirds. Although a few examples of specialized relationships between particular sunbirds and their flowers are known (e.g., golden-winged sunbird with shrubs of *Crotalaria* [Fabaceae] and *Leonotis* [Lamiaceae] and purple-breasted sunbird and *Symphonia globulifera*), these are exceptions rather than the rule in these birds. Although they are known to be aggressive defenders of patches of flowers and are very mobile after their breeding seasons, their effect on the genetic structure of populations of their food plants has apparently not yet been studied in much detail. But for areas such as eastern and southern Africa, which supports a high diversity of aloes (a genus containing about four hundred species) most of which are bird pollinated, it is highly likely that the evolution of these plants has been strongly influenced by sunbird foraging behavior.

An exception to this is a study of genetic subdivision and speciation in two herbaceous species of *Streptocarpus* (Gesneriaceae) growing in montane regions of South Africa. One species was pollinated by long-tongued flies; the other was pollinated by the malachite sunbird. As we might predict, the fly-pollinated species displayed greater between-population genetic differentiation and belonged to a species complex containing several recently evolved taxa. In contrast, the sunbird-pollinated species displayed lower population genetic differentiation and belonged to a clade containing lower species richness and relatively older taxa. These differences suggest that speciation rates have been higher in the fly-pollinated clade than in the sunbird-pollinated clade. Genetic evidence further indicated that pollen flow was more important than seed dispersal in determining genetic connections between populations in both species.

Much more research has been conducted on the genetic consequences of pollination by honeyeaters for their food plants, particularly in species-rich southwestern Australia—a region that contains many bird-pollinated shrubs and trees and many honeyeaters. Various studies of eucalypt trees and their avian pollinators, for instance, have shown that compared with insect flower visitors, honeyeaters deposit a greater number of different pollen genotypes on plant stigmas and produce higher rates of outcrossing (i.e., matings between different plants both within and between populations rather than self-fertilization). Nectar-feeding birds in Australia are typically wide ranging and presumably disperse pollen widely, which results in large effective population sizes and low genetic subdivision compared with wind- or insect-pollinated plants. Results of a study of the genetic structure of three species of eucalypts in southern Australia support this hypothesis. Compared with the honeyeater-pollinated species, fragmented populations of the two insect-pollinated species exhibited higher rates of self-fertilization and reduced genetic diversity.

Finally, although it is known that lorikeets feed in a very different fashion than other nectar-feeding birds, their effect on the population genetic structure of their food plants, particularly eucalypts, has seldom been studied. Their feeding behavior includes individuals "grazing" on many flowers per tree in the midst of a noisy flock of conspecifics and other species. These are very mobile birds, so we should expect them to be relatively long-distance pollen dispersers like other mobile nectar-feeding Australian birds and bats. Genetic data for blue gum trees, for instance, indicates that the very wide-ranging swift parrot, which is not a lorikeet, sometimes carries pollen from mainland Victoria to eastern Tasmania, a distance of over 500 km.

My overall conclusion here is that although we know a lot about the natural history of old-world nectar-feeding birds and bats, we still know relatively little about the genetic consequences of their foraging behavior compared with their new-world counterparts. Nonetheless, it is highly likely that, like many new-world nectar feeders, their food choices and foraging behavior have significant population genetic consequences in terms of levels of inbreeding, loss of genetic diversity, and genetic connections between isolated populations. Because of their conservation implications, these kinds of studies should have a high research priority.

7

CONSERVATION OF THESE MUTUALISMS

In this book I have sought to convince you that nectar-feeding birds and bats around the world have played and continue to play a significant role in the evolution of certain aspects of life on Earth, especially during the last twenty to twenty-five million years. There's no question that many avian nectar feeders are spectacularly beautiful, and many have evolved amazing lifestyles. But they are more than just a bunch of "pretty faces." Their role as "mobile links" in the economy of tropical, subtropical, and in some cases temperate ecosystems is also beyond question. By *mobile links*, which is a term the evolutionary biologist and butterfly expert Larry Gilbert coined to refer to the role that species of pollinators and seed dispersers play in ecosystems, I mean species whose movements connect populations ecologically (via seed dispersal) and genetically (via pollination). Vertebrate pollinators worldwide are very mobile animals that often acquire pollen on their beaks, snouts, feathers, and fur when they feed. Whenever their search for food takes them to new populations or habitats, they can potentially introduce new genes into them. By doing this, they are linking plant populations together in an otherwise increasingly fragmented world. They are also potentially rescuing plants from extinction by increasing the genetic diversity within their populations, thereby increasing their evolutionary resilience in a changing world. The bottom line here is that the world cannot afford to lose its vertebrate pollinators just as it cannot afford to lose the insects that pollinate our food crops.

So what is the current conservation status of our world's vertebrate nectar feeders? How many of these species of birds and bats are endangered or at risk of extinction, primarily as the result of the man's activities? The International Union for Conservation of Nature (IUCN) is an important organization that attempts to regularly assess the conservation status of as many of Earth's species of plants and animals as possible. The good news is that because many birds and mammals are conspicuous and charismatic, the status of many of their species is quite well known.

The current IUCN assessment includes six different categories of risk: LC = least concern; NT = near threatened; VU = vulnerable; EN = endangered; CR = critically endangered; and EW = extinct in the wild. The good news for three of the four families of nectar-feeding birds that I have discussed in this book is that over 75 percent of their species are classified as LC and very few species are VU (figure 26). As is well known, parrots worldwide are of considerable conservation concern largely because of illegal trapping for the pet trade. A major threat to all of these birds is habitat destruction, as is true for virtually all of Earth's wildlife. Habitat fragmentation that results in small population sizes is always a threat to wildlife because it increases their risk of extinction. In addition, some species of hummingbirds are extinction-prone because of their small geographic ranges that are often restricted to single Andean valleys or mountaintops. Furthermore, many island-dwelling, plant-visiting species of birds and bats have suffered substantial population losses through human exploitation and introduced predators, including rodents and snakes.

A similar situation prevails for nectar-feeding bats. Among fifty-six species of glossophagine and lonchophylline phyllostomids, for instance, four species are classified as NT, three species as VU, three species as EN, and one species (an island dweller) as CR. The other species (80 percent) are classified as LC. Even though many of them are cave dwellers, they are not harvested for food in Latin America. But they are still at risk whenever indiscriminate killing of cave bats occurs because of misguided "vampire control" efforts. Only two of the six genera (with three species) of nectar-feeding pteropodids are likely to be classified as VU because they are cave dwellers that can be easily captured for food. The other solitary-roosting species are much less vulnerable to this kind of threat. Nonetheless, because they live in large foliage-roosting colonies and eat fruit in addition to nectar and pollen, large flying foxes (*Pteropus* and its relatives) are also highly threated because they are often classified as vermin

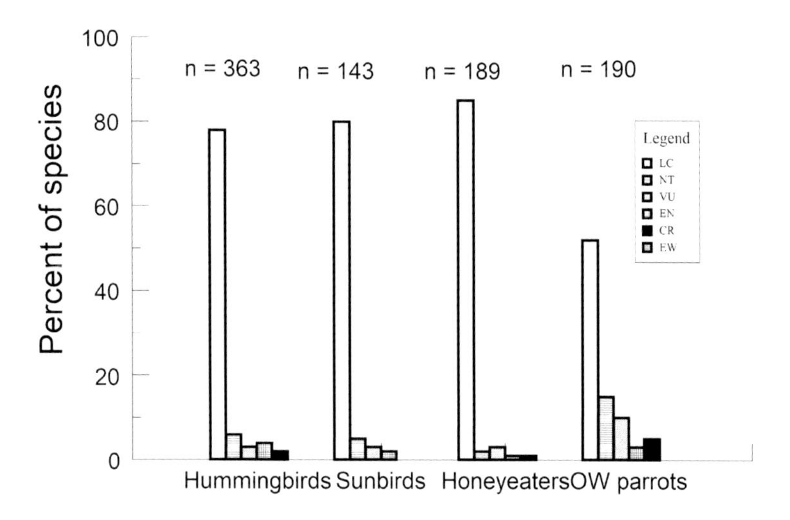

FIGURE 26 The conservation status of four groups of birds based on IUCN data as reported in https://birdsoftheworld.org. See the text for an explanation of the legend, whose symbols range from species of least concern (LC) to species extinct in the wild (EW).

and are killed near fruit orchards. They have also been killed for food on many Pacific Ocean islands as well as on the Asian mainland.

Although the ecological health of many vertebrate pollinators appears to be good now, we can still ask, how much would their food plants suffer if most of them were to go extinct? As an example of this approach, my colleagues and I conducted a series of pollinator exclusion experiments over several years for three species of Sonoran Desert vertebrate-pollinated columnar cacti—cardón, saguaro, and organ pipe. In these experiments we compared percent fruit-set resulting from flower visits by nocturnal (bats) and diurnal pollinators (mostly birds) with percent fruit-set in flowers visited by both kinds of pollinators (the controls). Results indicated that if bats were to disappear, percent fruit-set ranged from 12.5 percent (i.e., an 88 percent reduction) in cardón to 55.4 percent (a 45 percent reduction) in saguaro compared with controls. If birds were to disappear, percent fruit-set ranged from 10.4 percent (about a 90 percent reduction) in saguaro to 46.9 percent (a 53 percent reduction) in cardón. These results clearly indicate that bats are the most important pollinators of cardón whereas birds are the most important pollinators of saguaro. In the third species, organ pipe, birds are somewhat more important pollinators than bats. Overall, the loss of either flower-visiting bats or birds would result in drastic reductions in fruit (and seed) set in two of these three Sonoran Desert cacti. Since these large plants play a central role in the overall ecology of this desert, any reduction in their reproductive success resulting from the loss of their nectar-feeding (and seed-dispersing) vertebrates would be significant.

A metanalysis (i.e., the statistical analysis of many separate studies) of available pollinator exclusion studies published in 2018 supports the results of our cactus study. Based on a total of twenty-three and eighty-four studies in which nectar-feeding bats or birds, respectively, were excluded from flowers, bat exclusion resulted in an 83 percent reduction in fruit-set compared with a 46 percent reduction resulting from bird exclusion. Although a complete loss of vertebrate pollinators has not yet been reported for any plant species, these results indicate that plants would suffer a substantial reduction in their reproductive success if their vertebrate pollinators were to disappear. Island systems, mainly in the Pacific Ocean, are especially vulnerable to loss of vertebrate pollinators and seed dispersers, including pteropodid bats, fruit-eating pigeons, lorikeets, and Hawaiian honeycreepers, either directly or indirectly (i.e., by the introduction of invasive predators or competitors) by the hand of man. These losses will surely lead to a significant change in the ecology of these systems.

In thinking about the conservation of vertebrate pollinators, which as we've seen are most species-rich in the tropics, it is important to consider the structure of their communities and recognize that not all pollinators have the same impact on their plant communities. As is true of many communities of vertebrates, generalist pollinators with broad diets tend to far outnumber specialists in most of these communities. An effective conservation strategy for maximizing plant reproductive success should therefore be aimed at protecting generalists rather than just focusing on specialists. In addition to differences in diet breadth, vertebrate pollinators differ in many other ways, including their propensity to migrate either among local habitats or altitudinally and latitudinally. For example, many individuals of the lesser long-nosed bat, which feed at flowers of a variety of tropical dry forest trees and columnar cacti and agaves in more arid habitats, are latitudinal migrants in Mexico. These bats need intact populations of columnar cacti along Mexico's west coast in the spring and montane agaves in the fall to fuel their northern and southern migrations, respectively. Conservation of these food plants is therefore critical for maintaining their annual migratory cycle.

As a further example of the critical importance of conserving both vertebrate pollinators and their food plants for healthy ecosystems, let's consider the situation in the cape region of South Africa. The flora of this region is very rich, but the diversity of its major avian pollinators (only four species) is surprisingly poor. Nonetheless, it is believed that these birds have played a major role in the evolution of this flora. Evidence for this is that many of its prominent species are bird pollinated. Most of the nectar for birds here is produced by a profusion of shrubs of the protea or banksia family (Proteaceae), a family that also has many bird- (and bat-) pollinated species in Australia. Their colorful flowers often occur in cone-shaped, shaving brush–type inflorescences (figure 20). About 25 percent of its 330 species (in three genera—*Mimetes, Leucospermum,* and *Protea*) in this fire-maintained flora are pollinated by birds; the other species are insect pollinated. A statistical analysis of data from small plots and local landscapes indicates that the species richness and abundance of avian pollinators is more strongly correlated with the amount of nectar available in an area than with its vegetation structure (e.g., the density of shrubs) or postfire age of the vegetation. Most of this nectar is produced in the winter by species of *Protea,* which is when the nectar-feeding birds breed. But at times of low *Protea* flowering (e.g., in the summer), nectar produced by flowers of *Mimetes* and *Leucospermum* also becomes important for these birds. Consequently, conservation of species of all three plant taxa is critical for maintaining intact communities of avian pollinators in this region. Since most of these plants are self-incompatible (i.e., they cannot use their own pollen to fertilize their ovules), they are likewise critically dependent on abundant bird pollinators for their reproductive success. The fact that some of these plants are species of conservation concern, particularly species of *Mimetes* and *Leucospermum,* emphasizes the danger that this mutualistic system faces in a future that will include more frequent fires resulting from climate change.

In summary, the plant-animal mutualisms that I've described in detail in this book represent a kind of two-edged sword. On the positive side, the ecological and evolutionary interactions between flowering plants and their vertebrate pollinators have resulted in an increase in the overall beauty (to us) of many of Earth's ecosystems and have definitely produced a notable increase in Earth's overall biodiversity. After all, much of the rise of flowering plants beginning in the Late Cretaceous occurred as a result of their interactions with animal pollinators and seed dispersers. This increased diversity is especially impressive when you also consider all the nonvertebrate pollinators and their flowers that

I haven't treated in this book. As I write this, for example, the Sonoran Desert around Tucson is bursting with a bumper crop of flowers—whites, yellows, oranges, purples, and deep blues—produced by insect-pollinated annual and perennial plants. And the tips of tall whip-like ocotillo plants are producing their red tubular hummingbird-pollinated flowers. A little later saguaro cacti will begin to flower, and white-winged doves and lesser long-nosed bats will be avidly visiting them. Pollinator activity is certainly at its peak in the desert at this time of the year. But what if many of these pollinators—invertebrates as well as vertebrates—were to disappear? If this were to occur, then, to paraphrase Rachel Carson, this beautiful habitat would begin to experience its own version of a "silent spring."

On the negative side, many of Earth's mutualisms are in a precarious position because of their vulnerability to various extinction pressures. Whenever mutualistic partners become too interdependent, the loss of one or more partners can jeopardize the continued existence of the remaining partners. As a result, many mutualistic systems should probably be considered to be "fragile" from a long-term point of view. That is, they might be viewed as being inherently extinction-prone, particularly in a rapidly changing world. This is especially true because of the rate at which our species is changing the composition of all of Earth's ecosystems and its climate. But on the bright side, perhaps mutualisms are not any more extinction-prone than other kinds of ecological relationships that have evolved here on Earth. Predator-prey interactions, for example, are also likely to collapse whenever one or more key interactors on either side disappears. Just look at what happened to plants and their herbivores when wolves were removed from the Yellowstone ecosystem. So always remember what the great conservation biologist Aldo Leopold (1887–1948) has told us: "The first rule of intelligent tinkering is to keep all the pieces!" We're clearly tinkering with virtually all of Earth's ecosystems and have seemingly paid little heed to Leopold's rule. I wonder whether this is a recipe for ecological disaster.

8

A FINAL WRAP-UP

In this book I have endeavored to take you on a grand journey about the evolution and coevolution of nectar-feeding birds and bats that have become important pollinators of many species of flowering plants, especially in the tropics. The plant-animal interactions I've described here began with the evolution of flowering plants at least 130 million years ago. Various kinds of insects were the initial pollinators of angiosperm flowers. But by early in the Cenozoic era (at least 30–50 million years ago), however, a variety of different birds had begun to visit flowers for nutrition and bats in two families began to do this somewhat more recently. By the middle of the Miocene epoch (about 15 million years ago), specialized nectar-feeding birds and bats were fully coevolving with flowers produced by many families of advanced angiosperms throughout the tropics and subtropics. Interestingly, this coevolution has played out somewhat differently in the New and Old Worlds. As a result, Nathan Muchhala and I have proposed that there now exist three different nectar-feeding vertebrate "worlds": (1) the neotropics where birds and bats are small; they hover at flowers; and they mostly visit tubular or cup-shaped flowers; (2) the Afrotropics and southern Asia where birds and bats are larger; they don't usually hover; and the birds have more specialized interactions with their flowers than generalist-feeding bats; and (3) Southeast Asia and Australasia where birds and bats are also larger; they don't hover; and both birds and bats have relatively generalized interactions with their flowers. These biogeographic differences represent a west to east gradient of feeding specialization across the world's tropics. Feeding specialization is highest in the neotropics and is lowest in Southeast Asia and Australasia. I have suggested that this gradient reflects significant biogeographic differences in the spatiotemporal predictability (STP) of floral resources: high in the neotropics and much lower in Australasia. Nonetheless, conservation of all of these vertebrates is critical for the reproductive success of their food plants. Because many nectar-feeding birds and bats can be highly mobile, they are also critical for promoting gene flow between plant populations in our increasingly fragmented world. Finally, our planet certainly contains many forms of multicellular stardust, but to me, none are more interesting and spectacular than nectar-feeding birds and bats and their flowers. Their evolution and coevolution are wonderful examples of how opportunistic life has been on planet Earth. They also remind me of the song: "Woo, woo, woo, what a little stardust can do-oo-oo."

Finally, as sentient (but not always wise) organic beings, we humans have the privilege of admiring and studying Earth's vertebrate nectar feeders. But we also have the responsibility for making sure that we protect them from our destructive tendencies. A world without hummingbirds, nectar bats, and their flowers would be a very sad place indeed.

ACKNOWLEDGMENTS

I THANK THE FOLLOWING PEOPLE for discussion, encouragement, and providing me with materials for this book: Doug Altshuler, Alex Badyev, Judie Bronstein, Steve Buchmann, Steve Dyer, Brock Fenton, Curt Freese, Tigga Kingston, John Kress, Jeff Ollerton, Paul Racey, Alejandro Rico-Guevara, Armando Rodríguez, Judith Shangold, Gary Stiles, Marco Tschpaka, and Dan Weisz. And as always, my wife Marcia. I also wish to thank Bruce Taubert for introducing me to the world of hummingbird photography using multi-flash techniques. Over the years, my studies of fruit- and nectar-feeding bats have been supported by the Smithsonian Institution, the U.S. National Science Foundation, National Geographic Society, the Fulbright Foundation, the Ted Turner Endangered Species Fund, the U.S. National Fish and Wildlife Foundation, the Arizona Game and Fish Department, Earthwatch, and the Universities of Missouri–St. Louis and Miami. Mil gracias a todos!

GLOSSARY OF SCIENTIFIC NAMES

Birds

ANDEAN GIANT HUMMINGBIRD *Patagona gigas*

ANDEAN HILLSTAR HUMMINGBIRD *Oreotrochilus sp.*

ANNA'S HUMMINGBIRD *Calypte anna*

ANTILLEAN MANGO HUMMINGBIRD *Anthracothorax dominicus*

BEE HUMMINGBIRD *Mellisuga helenae*

BLACK-BELLIED THORNTAIL HUMMINGBIRD *Discosura langsdorffi*

BOOTED RACKET-TAIL HUMMINGBIRD *Ocreatus underwoodii*

BROAD-BILLED HUMMINGBIRD *Cynanthus latirostris*

BROAD-TAILED HUMMINGBIRD *Selasphorus platycercus*

CALLIOPE HUMMINGBIRD *Selasphorus calliope*

CHIMNEY SWIFT *Chaetura pelagica*

COSTA'S HUMMINGBIRD *Calypte costae*

ECUADORIAN HILLSTAR HUMMINGBIRD *Oreotrochilus chimborazo*

FIERY TOPAZ HUMMINGBIRD *Topaza pyra*

GOLDEN-WINGED SUNBIRD *Drepanorhynchus reichenowi*

GREEN-BACKED FIRECROWN HUMMINGBIRD *Sephanoides sephanoides*

GREEN MANGO HUMMINGBIRD *Anthracothorax viridis*

HOUSE FINCH *Haemorhous mexicanus*

HUNTER'S SUNBIRD *Chalcomitra hunteri*

LEWIN'S HONEYEATER *Meliphaga lewinii*

LONG-BILLED HERMIT HUMMINGBIRD *Phaethornis longirostris*
LONG-TAILED HERMIT HUMMINGBIRD *Phaethornis superciliosus*
MALACHITE SUNBIRD *Nectarinia famosa*
MARVELOUS SPATULETAIL HUMMINGBIRD *Loddigesia mirabilis*
MOUNTAIN AVOCETBILL HUMMINGBIRD *Opisthoprora euryptera*
NOISY FRIARBIRD *Philemon conrniculatus*
PUERTO RICAN EMERALD HUMMINGBIRD *Chlorostilbon maugaeus*
PURPLE-BREASTED SUNBIRD *Nectarinia purpurieventris*
PURPLE-THROATED CARIB HUMMINGBIRD *Eulampis jugularis*
RAINBOW LORIKEET *Trichoglossus moluccanus*
RED-HEADED HONEYEATER *Myzomela erythrocephala*
RUFOUS HUMMINGBIRD *Selasphorus rufus*
SCINTILLANT HUMMINGBIRD *Selasphorus scintilla*
SICKLEBILL HUMMINGBIRD *Eutoxeres aquila*
SWALLOW-TAILED HUMMINGBIRD *Eupetomena macroura*
SWIFT PARROT *Lathamus discolor*
SWORD-BILLED HUMMINGBIRD *Ensifera ensifera*
THORNBILL HUMMINGBIRD *Ramphomicron dorsale*
WEDGE-TAILED SABREWING HUMMINGBIRD *Campylopterus curvipennis*
WHITE-NECKED JACOBIN HUMMINGBIRD *Florisuga mellivora*
WHITE-VENTED VIOLET-EAR HUMMINGBIRD *Colibri serrirostris*
WIRE-CRESTED THORNTAIL HUMMINGBIRD *Discosura popelairii*

Bats

BANANA BAT *Musonycteris harrisoni*
BIG BROWN BAT *Eptesicus fuscus*
BLACK FLYING FOX *Pteropus alecto*
BUFFY FLOWER BAT *Erophylla sezekorni*
CALIFORNIA LEAF-NOSED BAT *Macrotus californicus*
COMMISSARIS'S LONG-TONGUED BAT *Glossophaga commissarisi*
COMMON LONG-TONGUED BAT *Glossophaga soricina*
DARK LONG-TONGUED BAT *Lichonycteris obscura*
DAWN NECTAR BAT *Eonycteris spelaea*
GOLDMAN'S NECTAR BAT *Lonchophylla robusta*
GREATER SPEAR-NOSED BAT *Phyllostomus hastatus*
JAMAICAN FRUIT BAT *Artibeus jamaicensis*
LESCHENAULT'S ROUSETTE *Rousettus leschenaultii*

LESSER LONG-NOSED BAT *Leptonycteris yerbabuenae*
LITTLE RED FLYING FOX *Pteropus scapulatus*
LONG-TONGUED BLOSSOM BAT *Macroglossus minimus*
MEXICAN LONG-TONGUED BAT *Choeronycteris mexicana*
NORTHERN BLOSSOM BAT *Macroglossus minimus*
PALE SPEAR-NOSED BAT *Phyllostomus discolor*
PALLID BAT *Antrozous pallidus*
QUEENSLAND BLOSSOM BAT *Syconycteris australis*
SEBA'S SHORT-TAILED BAT *Carollia perspicillata*
SHORT-NOSED FRUIT BATS *Cynopterus spp.*
SILKY SHORT-TAILED BAT *Carollia brevicauda*
SPECTRAL BAT *Vampyrum spectrum*
TAILED TAILLESS BAT *Anoura caudifer*
TUBE-LIPPED BAT *Anoura fistulata*
UNDERWOOD'S LONG-TONGUED BAT *Hylonycteris underwoodi*

Plants

BALSA TREE *Ochroma lagopus*
BLUE GUM EUCALYPTUS *Eucalyptus globulus*
BRITTLEBUSH *Encelia farinosa*
BUMPY SATINASH TREE *Syzygium cormiflorum*
CALABASH TREE *Crescentia cujete*
CAPE HONEYSUCKLE *Tecoma capensis*
CARDÓN CACTI *Pachycereus pringlei*
DAISY FLEABANE *Erigeron karvinskianus*
DESERT BLUEBELLS *Phacelia campanularia*
DESERT MARIGOLDS *Baileya multiradiata*
FAIRY DUSTERS *Calliandra eriophylla*
INDIAN PAINTBRUSH *Castilleja linariifolia*
JIMSONWEED *Datura stramonium*
MESQUITE *Prosopis velutina*
MEXICAN POPPIES *Argemone mexicana*
OCOTILLO *Fouquieria splendens*
ORGAN PIPE CACTI *Stenocereus thurberi*
PALO VERDE TREE *Parkinsonia microphyllum*
PERUVIAN APPLE CACTUS *Cereus peruvianus*
PLANTAIN *Musa balbisiana*

POCHOTA TREE *Pochota (Bombacopsis) fendleri*

SAGUARO CACTI *Carnegiea gigantea*

SAUSAGE TREE *Kigelia africana*

SILK COTTON TREE *Ceiba pentandra*

SNAPDRAGON *Antirrhinum spp.*

TEXAS EBONY TREE *Ebenopsis ebano*

WATERMEAL *Wolffia spp.*

WEST INDIAN LOCUST TREE *Hymenaea courbaril*

NOTES

Note: Full citations of books and articles that I cite here can be found in the Bibliography.

1. Beginnings

Atoms, Earth History, and the Evolution of Life

I have used Wikipedia accounts for general overviews of many of the topics in this section before switching to other sources for most of the information that I present here and throughout this book. I realize that some people do not recommend using Wikipedia for information, but I disagree with this. I've consulted several colleagues on this issue, and we all agree that Wikipedia accounts of particular scientific topics are generally accurate; they are at least as accurate as similar accounts in *Encyclopedia Britannica* according to a survey published in *Nature*. As a result, I have no qualms about using Wikipedia to learn about new topics that are not in my intellectual toolkit. I hope you feel the same way.

Any recent book on cosmology (e.g., Perlov and Vilenkin [2017]) will contain a detailed discussion of the evolution of our universe. It's a mind-boggling story full of mathematical surprises (e.g., why is the discrepancy between the observed value of the cosmological constant and its value predicted by quantum field theory so large [by a factor of 50–120 orders of magnitude!]?). The history of the discovery of the nature of atoms is beautifully described in great detail by Rhodes (1986). Schrödinger's (1944) book sets the stage for my discussion of the evolution of life, a topic that is covered in detail in general biology and evolution textbooks (e.g., Freeman et al. [2014] and Zimmer and Emlen [2013]), respectively. Howard Markel's book *The Secret of Life* (2021) contains a fascinating account of the Watson-Crick (and Franklin) pursuit of the structure of DNA. The history of planet Earth is discussed in many geology books. A highly readable introduction can be found in Knoll (2021), which contains many accessible as well as technical references to this subject. His chapter 4 ("Oxygen Earth") discusses the evolution of Earth's currently oxygen-rich atmosphere. Finally, an excellent treatment of the evolution of life and the role of horizontal gene transfer in it can be found in David Quammen's book *The Tangled Tree* (2018). Mayr and Wilder (2004) present fossil evidence that birds were visiting flowers fifty million years ago.

On to the Vertebrates

Treatments of this general topic can be found in many books, including the biology and evolution texts mentioned above as well as specific "ology" books such as those dealing with fish, amphibians and reptiles, birds, and mammals. Chapter 5 ("Animal Earth") in Knoll (2021) is another useful source of information and recent references dealing with this topic. As I indicate in *Sonoran Desert Journeys* (2022), I have always been fascinated with this aspect of the evolution of life on Earth.

Greening of the Earth

This broad topic is also standard fare in general biology, botany, and evolution texts as well as in chapter 6 ("Green Earth") in Knoll (2021). Because plants in their many evolutionary forms—from nonvascular mosses and bryophytes to single-celled algae and giant sequoia trees—represent the energy base of most aquatic and terrestrial ecosystems, their history is particularly important in the evolution of life on Earth. The evolution of photosynthesis—the conversion of electromagnetic energy into chemical energy—has to have been the most important "invention" for most of the life that we see around us. Reviews of this topic include Blankenship (2010), Cardona (2019), Martin, Bryant, and Beatty (2018), and Xiong et al. (2000).

2. Tree Thinking

The Basics

I discuss the tools evolutionary biologists have used use to create hypotheses about the evolutionary histories of particular groups of organisms in *Sonoran Desert Journeys*. Zimmer and Emlen (2013) and many others do likewise. David Quammen (2018) reviews the history of thinking behind the Tree of Life concept and offers new ideas about what this tree actually looks like.

The Evolutionary History of Flowering Plants

This topic is again covered in general biology and botany textbooks (e.g., Mauseth 2021). John Kress and I cover this topic in *The Ornaments of Life* (2013) as does Stephen Buchmann in *The Reason for Flowers* (2015). Soltis et al. (2018) contains a thorough treatment of this history from a phylogenetic perspective. Magallón et al. (2015) present a time-calibrated phylogeny of flowering plants. Charles Darwin considered their evolution to be an "abominable mystery" because of their sudden appearance in the fossil record, but if it hadn't happened, I wouldn't be writing this book.

The Evolutionary History of Birds

Ornithology texts (e.g., Gill and Prum 2019) cover this topic. Brusatte, O'Connor, and Jarvis (2015) describe the morphological changes that transformed certain dinosaurs into birds. Recent phylogenetic treatments of avian history include Jarvis et al. (2014), Claramunt and Craycraft (2015), Prum et al. (2015), and Oliveros et al. (2019). Voskamp et al. (2017) summarize global patterns of current avian diversity and its conservation.

The Evolutionary History of Mammals

Likewise, mammalogy texts (e.g., Feldhammer et al. 2020) cover this topic and describe the evolutionary transitions from synapsids to mammals. Recent phylogenetic treatments of mammalian history include Bininda-Emonds et al. (2007), O'Leary et al. (2013), and Upham, Esselstyn, and Jetz (2019). Geographic patterns of mammalian diversity are described by Ceballos and Brown (1995), Davies et al. (2008), and Safi et al. (2011), among others.

3. How to Build a Hummingbird

An Overview of Hummingbird Diversity and Evolution

Schuchmann (1999) and McGuire et al. (2014) review and discuss the diversity and evolution of hummingbirds, respectively. Leimberger et al. (2022) discuss hummingbird evolution from a plant-animal interaction perspective. Weinstein et al. (2014) review hummingbird morphological diversity, and Bleiweiss (1998) discusses the origins of hummingbird faunas. Sonne et al. (2016) report on the unusually small geographic range sizes in these birds.

The Emergence of Swifts and Hummingbirds

I used the following sources in my discussion of the evolution of the avian order Strisores and the swift and hummingbird families: Bleiweiss (1998), Chen et al. (2019), Mayr (2003, 2004), McGuire et al. (2014), and White and Braun (2019). Lovette, Bermingham, and Ricklefs (2002) discuss the evolution of Hawaiian passerines, and Traveset et al. (2015) describe the birds pollinating flowers in the Galapagos. Navalón et al. (2020, 2021) conducted detailed analyses of the evolution of the skulls of Strisores and island-dwelling passerines. Ashton (2014, chapters 5 and 6) and Zhao, Hou, and Li (2022) set the stage for the emergence of many groups of modern plants, birds, and mammals in Eurasia. Information about the genetic basis for the evolution of some of the morphological features of vertebrates comes from Freeman et al. (2014, chapter 22), Shubin, Tabin, and Carroll (1997), and Zimmer and Emlen (2013, chapter 10).

A Comparison of Modern Hummingbirds and Swifts or How Different Have They Become in Forty-Two Million Years?

Summaries of the basic characteristics of these two families can be found in any bird encyclopedia such as the older Perrins and Middleton (1985) and the website https://birdsoftheworld.org administered by the Cornell Laboratory of Ornithology. Wikipedia accounts, particularly those of hummingbirds, also provide very useful overviews of these families. The Navalón et al. studies cited above involve morphological comparisons of members of the Strisores.

The Basic Biological and Ecological Characteristics of Hummingbirds

Why Is Their Small Size So Important? Calder (1984) is a classic book dealing in detail with this subject. I discuss some of the consequences of body size in *Sonoran Desert Journeys*.

The Consequences of Being an Endotherm: McNab (2002, 2012) discusses this topic in great deal. I present an abbreviated account of it in *Sonoran Desert Journeys*.

How High Are the Metabolic Rates of Hummingbirds Compared with Other Birds and Mammals? There is a large literature on this subject, and I will cite only a few articles here. These include studies of basic metabolism (Osipova et al. [2023], Suarez [1992, 2013], Suarez and Welch [2017], and Workman et al. [2018]); field metabolic rates (Nagy, Girard, and Brown [1999] and Powers and Nagy [1988]); hyperglycemia (Beuchat and Chong [1998] and Hargrove [2005]); and torpor (Ruf and Geiser [2015], Spence and Tingley [2021], Wellbrock et al. [2022], and Wolf et al. [2020]). The metabolism of swifts is discussed by Doucette et al. (2012), McNab (1988), and McNab and Bonaccorso (1995).

How Hummingbirds Meet Their Oxygen Needs: Papers dealing with this topic include Chai and Dudley (1996), Suarez (1998), and Welch, Altshuler, and Suarez (2007).

The Challenges of Living at High Elevations: Many hummingbirds thrive at high elevations. How they do this and its costs are discussed by Altshuler (2006), Altshuler and Dudley (2002), Altshuler, Dudley, and McGuire (2004), Lim et al. (2019), Projecto-Garcia et al. (2013), Stiles (2004, 2008), and Wolf et al. (2020).

Hummingbird Wings and How They Hover: This unique flight ability has been much studied. Classic work includes Greenewalt (1960). Recent papers include Altshuler et al. (2010), Goller and Altshuler (2014), Groom, Toledo, and Welch (2017), Ingersoll et al. (2018), Kruyt et al. (2014), Maeda et al. (2017), Sapir and Dudley (2012, 2013), Shankar et al. (2019), Skandalis et al. (2017), and Warrick et al. 2012.

Hummingbird Feathers: All That Glitters . . . : Greenewalt (1960) was instrumental in determining how iridescence is produced by hummingbird feathers. The basic principles are nicely illustrated by Bartley and Swash (2022). Recent papers dealing with this subject include Clark, Elias, and Prum (2011), D'Alba et al. (2021), Diamant, Falk, and Rubenstein (2021), and Venable, Gham, and Prum (2022).

A Tale of Hummingbird Tails: The diversity and evolution of hummingbird tails is discussed by Bleiweiss (2009), Clark (2010), Clark and Dudley (2009), Clark, Elias, and Prum (2011), Johnsgard (2016), Thomas (1997), and Nan et al. (2020). This diversity can be seen in any neotropical bird guide (e.g., Schulenberg et al. 2007).

How Hummingbirds Harvest Food: Hummingbird bills and tongues have attracted much research attention recently. Many books (e.g., Bartley and Swash 2022) portray the diversity of their bills. Sexual dimorphism in bills is discussed by Berns and Adams (2013), Bleiweiss (1999), and Temeles, Miller, and Rifkin (2010), among others. Trait matching between hummingbird bills and their flowers is discussed by Dalsgaard et al. (2021), Rico-Guevara et al. (2021), and Weinstein and Graham (2017). How tongues work is described by Rico-Guevara, Rubega, et al. (2017) and Rico-Guevara et al. (2018, 2023). And bills as weapons was reported by Rico-Guevara and Araya-Salas (2015). Yanega and Rubega (20024) describe bill flexion in these birds. The insect diets of hummingbirds and swifts are discussed by Collins (2015), Moran, Prosser, and Moran (2019), Nocera et al. (2012), Orlowski and Karg (2013), Stiles (1995), and Warrick et al. (2012).

Water, Water Everywhere: The Water Economy of Hummingbirds: Because of their diet and small size, hummingbirds are faced with a unique situation regarding their water balance. Papers discussing this include Bakken and Sabat (2008), Beuchat, Calder, and Braun (1990), Lotz and Martínez del Rio (2004), and McWhorter and Martínez del Rio (1999).

How Do Hummingbirds Perceive Their World? Ed Yong (2022) discusses the many ways that animals perceive their world. Hummingbirds are very visual animals. Their visual adaptations are discussed by Altshuler and Wiley (2020), Gill and Prum (2019), Goller and Altshuler (2014), Herrera et al. (2008), Ibbotson (2017), Tyrell et al. (2018), and White et al. (2021).

Hearing and Vocalizations: Although they are not passerine songbirds, hummingbirds are very vocal. Papers dealing with their hearing and song learning include Araya-Salas et al. (2019), Clark et al. (2018), Duque and Carruth (2022), Duque et al. (2020), Gaunt et al. (1994), and Riede and Olson (2020).

Brains and Sound Reception: The following papers deal with hummingbird brains and sound reception: Baptista and Schuchmann (1999), Jarvis (2004), Jarvis et al. (2000), Jarvis and Consortium (2005), Ocampo, Barrantes, and Uy (2018), and Ricklefs (2004).

More on Vocal Learning: In addition to papers cited above, the following papers continue with this discussion: Baptista and Schuchmann (1999), Gahr (2000), Johnson and Clark (2020), Monte, da Silva, and Gahr (2023), Nowicki and Searcy (2014), and ten Cate (2021)

Putting It All Together, Part 1: Voices, Tails, and Wings as Communication "Instruments": This section is based on papers cited above under the appropriate topics; see especially Clark et al. (2018). In addition, discussion of sexual selection in these birds can be found in Falk, Webster, and Rubenstein (2021), and Temeles, Miller, and Rifkin (2010).

Putting It All Together, Part 2: The Cognitive Abilities of Hummingbirds: Their cognitive abilities have been studied explicitly by Araya-Salas et al. (2018) and González-Gómez and Araya-Salas (2019).

Additional Hummingbird Senses: Smell and Taste: These important senses are discussed in the following papers: Baldwin et al. (2014), Cockburn et al. (2014), Ioale and Papi (1989), Kim, Rankin, and Wilson Rankin (2021), Medina-Tapia et al. (2012), Nuñez, Méndez, and López-Rull (2021), Rankin, Clark, and Wilson Rankin (2018), and Wang et al. (2019). Floral scents of hummingbird flowers are discussed by Knudsen et al. (2004).

Hummingbird Foraging Behavior: Optimal or Not? Many papers deal with hummingbird foraging behavior. Discussions of optimal foraging theory include Pyke (1984, 2016) and Schoener (1971). Applications of this concept to hummingbirds and descriptions of their foraging behavior can be found in Bateson, Healy, and Hurley (2003), Baum and Grant (2001), Gill (1998), González-Gómez, Vásquez, and Bozinovic (2011), Hixon, Carpenter, and Paton (1983), and Mitchell (1989).

Migration: Hummingbirds on the Move: Many hummingbirds change habitats seasonally and/or migrate altitudinally or latitudinally. Papers discussing this topic include Boyle and Conway (2007), Eberts, Guglielmo, and Welch (2021), Levey and Stiles (1992), Rappole and Schuchmann (2003), Salewski and Bruderer (2007), Sol et al. (2010), and Zenzal et al. (2018).

How Hummingbirds Make Babies: Mating systems and reproduction are discussed in Bartley and Swash (2022), Bleiweiss (1999), Gonzalez and Ornelas (2014), Johnsgard (2016), Stiles (1980), and Stiles and Wolf (1979).

Hummingbirds in a Community Context: How Important Is Competition for Food? Many papers deal with this topic, including Abrahamczyk and Kessler (2015), Feinsinger and Colwell (1978), Graham et al. (2009), Kodric Brown et al. (1984), Lara (2006), Pigot and Etienne (2015), Rodríguez-Flores et al. (2019), and Weinstein and Graham (2017).

4. How to Build a Nectar Bat

An Overview of Nectar Bat Diversity and Evolution

Papers dealing with the evolution of bats include Bishop (2008), Rubalcaba et al. (2022), Simmons et al. (2008), Shen et al. (2010), Shi and Rabosky (2015), Teeling et al. (2005), and Zhang et al. (2013). The specific evolution of phyllostomid bats is reviewed by Datzmann, von Helversen, and Mayer (2010), Davalos, Velazco, and Rojas (2020), Davalos et al. (2020), Monteiro and Nogueira (2011), and Simmons, Gunnell, and Czaplewski (2020). Their current classification occurs in Cirranello and Simmons (2020). Overviews of the evolution of echolocation in bats can be found in Denzinger, Tschapka, and Schnitzler (2018), Fenton (2013), Jones and Teeling (2006), Kalko and Schnitzler (1998), Schnitzler and Kalko (1998), and Thiagavel et al. (2020).

Comparison of Nectar Bats and Their Ancestors

Giannnini and Velazco (2020) summarize the basic differences between mormoopid and phyllostomid bats. Muchhala and Tschapka (2020) discuss many aspects of the ecology and evolution of phyllostomid nectar bats.

Basic Biological and Ecological Characteristics of Nectar Bats

First Things First: The Morphological Nuts and Bolts of Being a Bat: Many aspects of the morphology of all bats differs profoundly from that of other terrestrial mammals. Their wings are obviously their defining feature, but many other parts of their bodies (e.g., their hind limbs) have become modified to accommodate their unique lifestyle. I like the detailed treatment of their morphology in Terry Vaughan's mammalogy textbooks (e.g., Vaughan, Ryan, and Czaplewski 2000) because he has studied their morphology in great detail throughout his career. His descriptions of differences in the functional morphology of mormoopid and phyllostomid bats, including nectar-feeding species, is outstanding. Other important papers dealing with phyllostomid morphology, especially their wings and skulls, include Bolzan et al. 2015, Freeman (1995, 2000), Gardiner, Codd, and Nudds (2011), Norberg and Rayner (1987), and Tavares (2013).

How Nectar Bats Hover: In addition to Vaughan et al. (2000), Ingersoll, Haizmann, and Lentink (2018) and Vejdani, Boerman, et al. (2019) treat their hovering behavior compared with hummingbirds in detail.

Metabolic Consequences of Size, Diet, and Hovering: Cruz-Neto and Herrera (2020) provide an overview of the basic metabolism and water processing of phyllostomid bats. Basic metabolic studies include Bonaccorso et al. (1992), McNab (1986, 2003, 2012), Shen et al. (2010), and Voigt and Speakman (2007). Studies of field metabolic rates include Voigt, Kelm, and Visser (2006). Hovering and its cost are described by Vejdani, Boerman, et al. (2019), Voigt and Winter (1999), and Welch, Myrka, et al. (2018). Topor in these and other bats is discussed by Kelm and von Helversen (2007) and Ruf and Geiser (2015). Sugar metabolism in nectar bats has been studied by Castro et al. (2021), Potter et al. (2021a), Suarez and Welch (2017), Suarez et al. (2009), and Welch and Chen (2014).

Water, Water Everywhere Redux: As indicated above, Cruz-Neto and Herrera (2020) provide an overview of the water economy of phyllostomid bats, including the nectar feeders. Carpenter (1969) specifically determined that lesser long-nosed bats produce extremely dilute urine. More recent papers on this topic include Potter et al. (2021b) and Schondube, Herrera, and Schondube (2001).

How Do Nectar Bats Perceive and Interact with Their World?

Bat Brains and Vision: The basic features of bat brains involved in food gathering are discussed by Arbour, Curtis, and Santana (2021), Carter, Ratcliffe, and Galef (2010), Dumont (2004), Henry and Stoner (2011), Ratcliffe (2009), Rojas et al. (2013), Safi and Dechmann (2005), and Thiagavel et al. (2020). Papers dealing with their vision include Bell and Fenton (1986), Gorreson et al. (2015), Hall et al. (2021), and Kries et al. (2018).

How Important Is Scent to These Bats? This topic is discussed by Brokaw and Smotherman (2021), Brokaw et al. (2021), Gonzalez-Terrazas et al. (2016a, 2016b), and Muchhala and Serrano (2015).

Spatial Memory: Papers or chapters discussing this important behavior include Carter, Ratcliffe, and Galef (2010), Henry and Stoner (2011), Muchhala and Tschapka (2020), and Tschapka (2004).

Putting It All Together: The experimental studies of Gonzalez-Terrazas and colleagues cited above describe the importance of echolocation and scent as lesser long-nosed bats approach and feed at large cactus flowers. Vision is undoubtedly also important in this behavioral task.

How Do Nectar Bats Feed? Muchhala and Tschapka (2020) review this topic in detail. Papers dealing with it include von Helversen and Reyer (1984), Lemke (1984), Fleming (2022), Horner, Fleming, and Sahley (1998), and Tschapka (2004).

How Do Nectar Bats Make Babies? The reproductive biology of phyllostomid bats is reviewed by Barclay and Fleming (2020). That of the nectar-feeding lesser long-nosed bat is described by Fleming and Martino (2020) and Fleming and Nassar (2002).

What's Their Gregarious Lifestyle Like? How Social Are Nectar Bats? The social behavior of phyllostomid bats is reviewed by Adams, Nicolay, and

Wilkinson (2020) and McCracken and Wilkinson (2000). Case studies of well-known species, including the lesser long-nosed bat, can be found in Fleming and Martino (2020). Life in a maternity colony of the lesser long-nosed bat is described by Fleming, Nelson, and Dalton (1998). Murray and Fleming (2008) describe the social structure of the buffy flower bat.

How Does the Community Structure of Nectar Bats Compare with That of Hummingbirds? Fleming, Muchhala, and Ornelas (2005) review this topic in great detail. Heithaus, Fleming, and Opler (1975) and Tschapka (2004) describe the nectar bat communities at La Pacifica and La Selva in Costa Rica, respectively.

5. Putting These Vertebrates and Their Food Plants Together

Phylogenetic Distribution of These Pollination Modes

This section is based on the comprehensive reviews found in Fleming and Kress (2013, chapter 7), Fleming and Muchhala (2008), and Fleming, Geiselman, and Kress (2009).

Evolutionary Transitions from Insect to Vertebrate Pollination

The phylogenetic summary comes from Fleming and Kress (2013, chapter 7). Many biologists, including Buchmann (2015), Endress (1994), and Proctor, Yeo, et al. (1996), have discussed this topic. Recent papers include Armbruster (2014), Endress (2011), Moyroud and Glover (2017), Ronse de Craene (2018), and Soltis et al. (2008).

How to Make a Flower

Floral Developmental Genetics: This topic is a very active field of research. A few recent reviews include Hileman (2014), Irish (2017), Mayroud and Glover (2017), and Soltis et al. (2018).

Other Floral Traits: Fleming and Kress (2013, chapter 7) review the evolution of acoustic features, color, phenology, presentation, and scent of vertebrate-pollinated flowers. Additional references include Amrad et al. (2016), Baker and Baker (1982), Buchmann (2015), CaraDona, Iler, and Inouye (2014), Forrest, Inouye, and Thomson (2010), Lee (2007), McKinney et al. (2012), Raguso (2008), Simon et al. (2023), and Stiles (1978). The pollination syndrome concept is reviewed by Dellinger (2020) and Rosas-Guerrero et al. (2014), among many others.

Genetic Consequences of Vertebrate Pollination

References that I used for this section include Abrahamczyk, Souto-Vilarós, and Renner (2014), Alexandre et al. (2015), Ashokan et al. (2022), Dunphy, Hamrick, and Swagerl (2004), Fleming (2006), Fleming and Kress (2013, chapter 4), Gamba and Muchhala (2021), Givnish et al. (2014), Hernández-Hernández et al. (2014), Lagomarsino et al. (2016), Lagomarsino et al. (2017), Lopes et al. (2022), Martén-Rodríguez, Almares-Castro, and Fenster (2009), Martén-Rodríguez et al. (2010), Mungia-Rosas et al. (2009), Quesada et al. (2004), Serrano-Serrano et al. (2017), Torres-Vanegas et al. (2019), and Wessinger, Rausher, and Hileman (2019).

What Is Coevolution?

Classic treatments of this topic include Futuyma and Slatkin (1983) and Janzen (1980). A broad treatment of this topic occurs in Fleming and Kress (2013, chapter 5). Abrahamcyzk, Poretschkin, and Renner (2017) examine this concept in five bird-plant associations.

Putting It All Together Again: Case Studies of the Evolution of Hummingbirds, Nectar Bats, and Their Flowers

References that I used for this section include Abrahamcyzk and Renner (2015), Abrahamcyzk, Poretschkin, and Renner (2017), Aguilar-Rodríguez et al. (2019), Fleming et al. (2001), Lagomarsino et al. (2017), Martén-Rodríguez, Almares-Castro, and Fenster (2009), Mungia-Rosas et al. (2009), Roalson and Roberts (2016), Tripp and Manos (2008), and Wessinger, Rausher, and Hileman (2019).

Wrapping It All Up

References in this section include Abrahamcyzk and Renner (2015), Osipova et al. (2023), and Wilkinson et al. (2021).

6. Paleotropical Nectar-Feeding Birds and Bats and Their Food Plants

Overview of Diversity, Distribution, and Evolution of These Vertebrates

Summaries of the diversity and biology of sunbirds, honeyeaters, and lorikeets can be found in Cheke, Mann, and Allen (2001), Perrins and Middleton

(1985), and https://birdsoftheworld.org. Hall and Richards (2000) and Taylor and Tuttle (2019) cover these topics for pteropodid nectar feeders. Specific articles include Fleming and Muchhala (2008), which contains an overview of their pollination ecology, and a series of papers dealing with their evolutionary histories, including Almeida, Simmons, and Giannini (2020), Driskell and Christidis (2004), Hassanin et al. (2020), Jarvis et al. (2014), Marki et al. (2019), Provost, Joseph, and Smith (2018), Schweizer et al. (2015), Smith et al. (2020), and Warren et al. (2003). Tim Low provides an especially good overview of the evolution and ecology of Australasian honeyeaters and lorikeets in *Where Song Began* (2016). Wester (2014) reviews evidence for hovering in old-world flower-visiting birds.

An Overview of Their Major Biological Features

Sunbirds and Honeyeaters: References here include Campillo et al. (2018), Cheke, Mann, and Allan (2001), Fleming and Nicolson (2003), Fleming, Gray, and Nicolson (2004), Ford and Paton (1977), Janacek et al. (2021), Lauron et al. (2014), Low (2016), Moyle et al. (2011), Nicolson and Fleming (2003), Nsor, Godsoe, and Chapman (2019), Recher and Davis (2011), Roxburg and Pinshow (2000), and Wooller, Richardson, and Pangendham (1988).

Flowerpeckers and Lorikeets: References here include Brice, Dahl, and Grau (1989), Jaggard et al. (2015), Joseph et al. (2014), Karasov and Cork (1996), Franklin and Noske (1999), Low (2016), Pyke (1980), Smith and Lill (2008), Stanford and Lill (2008), and Waterhouse (1997).

Nectar Bats: Papers and books discussing biology and evolution of these bats include Amitai et al. (2010), Anderson and Ruxton (2020), Bonaccorso and McNab (1997), Bonaccorso et al. (2002), Bumrungsri et al. (2013), Eby (1991), Hall and Richards (2000), Hodgkison et al. (2003), Law (1993), McNab and Bonaccorso (2001), Start and Marshall (1976), Stewart and Dudash (2018), Teeling (2009), Thiagavel et al. (2018), Veselka et al. (2010), Voigt et al. (2011), and Winkleman et al. (2003).

The Kinds of Flowers Pollinated by Old-World Birds and Bats

References in the first section of part 5 indicate the kinds of flowers that are visited by both new- and old-world vertebrate nectar feeders. Fleming and Kress (2013, chapter 7), Fleming and Muchhala (2008), and Fleming, Geiselman, and Kress (2009) provide overviews of this topic. Aloes in South Africa and eucalypts in Australia very important flower sources for birds and bats.

Discussions of their evolution can be found in Adams et al. (2000), Grace et al. (2013), Manning et al. (2014), and Thornhill et al. (2019). Trees of the genus *Syzygium* are also important flower (and fruit) resources for Australian birds and bats. Their evolution is discussed by Low et al. (2022).

Genetic Consequences of Paleotropical Vertebrate Pollination

References dealing with this topic include Bezemer et al. (2019a, 2019b), Crome and Irvine (1986), Hopper and Moran (1981), Hopper et al. (2021), Hughes et al. (2007), Law and Lean (1999), Sampson, Hopper, and James (1989, 1995), and Xiang et al. (2022).

7. Conservation of These Mutualisms

Information about the conservation status of nectar-feeding birds and bats comes from the IUCN Redlist (https://www.iucnredlist.org) and Birds of the World (https://birdsoftheworld.org). References for this section include Aguilar-Rodríguez et al. (2019), Bachmann et al. (2020), Buchmann and Nabhan (1996), Cox (1983), Fleming et al. (2001), Funamoto (2019), Geerts et al. (2020), Gilbert (1980), Jones et al. (2009), Leimberger et al. (2022), Menz et al. (2011), Nassar et al. (2020), Pauw (2019), Prieto-Torres et al. (2022), Ratto et al. (2018), Remolina-Figueroa et al. (2022), Wethington, West, and Carlson (2005), Wiles and Brooke (2009), and Zenata et al. (2017). Fleming and Kress (2013, chapter 10) discuss this topic in detail. Aldo Leopold's quote comes from *Round River* (1972).

8. A Final Wrap-up

The "three world view" of nectar-feeding vertebrates comes from Fleming and Muchhala (2008).

BIBLIOGRAPHY

Abrahamczyk, S. and M. Kessler (2015). "Morphological and behavioural adaptations to feed on nectar: how feeding ecology determines the diversity and composition of hummingbird communities." *Journal of Ornithology* 156: 333–347.

Abrahamczyk, S., C. Poretschkin, et al. (2017). "Evolutionary flexibility in five hummingbird/plant mutualistic systems: testing temporal and geographic matching." *Journal of Biogeography* 2017: 1–9.

Abrahamczyk, S. and S. S. Renner (2015). "The temporal build-up of hummingbird/plant mutualisms in North America and temperate South America." *Bmc Evolutionary Biology* 15. https://doi.org/10.1186/s12862-015-0388-z.

Abrahamczyk, S., D. Souto-Vilaros, et al. (2014). "Escape from extreme specialization: passionflowers, bat, and the sword-billed hummingbird." *Proceedings of the Royal Society B-Biological Sciences* 281: 1–7.

Adams, D. M., C. W. Nicolay, et al. (2020). Patterns of sexual dimorphism and mating systems. *Phyllostomid Bats: a Unique Mammalian Radiation*. T. H. Fleming, L. M. Davalos and M. A. R. Mello. Chicago, University of Chicago Press: 221–238.

Adams, S. P., I. J. Leitch, et al. (2000). "Ribosomal DNA evolution and phylogeny in *Aloe* (Asphodelaceae)." *American Journal of Botany* 87: 1578–1583.

Aguilar-Rodríguez, A., M. Tschapka, et al. (2019). "Bromeliads going batty: pollinator partitioning among sympatric chiropterophilous Bromeliacae." *Annals of Botany Plants* 2019: 1–19.

Alexandre, H., J. Vrignaud, et al. (2015). "Genetic architecture of pollination syndrome transition between hummingbird-specialist and generalist species in the genus *Rhytidophyllum* (Gesneriaceae)." *PeerJ*: 1–24.

Almeida, F. C., N. B. Simmons, et al. (2020). "A species-level phylogeny of Old World fruit bats with a new higher-level classification of the family Pteropodidae." *American Museum Novitates* (3950): 1–24.

Altshuler, D. L. (2006). "Flight performance and competitive displacement of hummingbirds across elevational gradients." *American Naturalist* 167: 216–229.

Altshuler, D. L. and R. Dudley (2002). "The ecological and evolutionary interface of hummingbird flight physiology." *Journal of Experimental Biology* 205(16): 2325–2336.

Altshuler, D. L., R. Dudley, et al. (2010). "Allometry of hummingbird lifting performance." *Journal of Experimental Biology* 213: 725–734.

Altshuler, D. L., R. Dudley, et al. (2004). "Resolution of a paradox: hummingbird flight at high elevation does not come without a cost." *Proceedings of the National Academy of Sciences of the United States of America* 101: 17731–17736.

Altshuler, D. L. and D. R. Wiley (2020). "Hummingbird vision." *Current Biology* 30: R95–R111.

Amitai, O., S. Holtze, et al. (2010). "Fruit bats (Pteropodidae) fuel their metabolism rapidly and directly with exogenous sugars." *Journal of Experimental Biology* 213(15): 2693–2699.

Amrad, A., M. Moser, et al. (2016). "Gain and loss of floral scent production through changes in structural genes during pollinator-mediated speciation." *Current Biology* 26: 3303–3312.

Anderson, S. C. and G. D. Ruxton (2020). "The evolution of flight in bats: a new hypothesis." *Mammal Review* 50: 426–439.

Araya-Salas, M., P. L. Gonzalez-Gomez, et al. (2018). "Spatial memory is as important as weapon and body size for territorial ownership in a lekking hummingbird." *Scientific Reports* 8: 1–11.

Araya-Salas, M., G. Smith-Vidaurre, et al. (2019). "Social group signatures in hummingbird displays provide evidence of co-occurrence of vocal and visual learning." *Proceedings of the Royal Society B-Biological Sciences* 286: 1–9.

Arbour, J. H., A. A. Curtis, et al. (2021). "Sensory adaptations reshaped intrinsic factors underlying morphological diversification in bats." *Bmc Biology* 19: 1–13.

Armbruster, W. S. (2014). "Floral specialization and angiosperm diversity: phenotypic divergence, fitness tradeoffs and realized pollination accuracy." *Annals of Botany Plants* 6: 1–24.

Ashokan, A., J. Leong-Skornickova, et al. (2022). "Floral evolution and pollinator diversification in *Hedychium*: revisiting Darwin's predictions using an integrative taxonomic approach." *American Journal of Botany* 2022: 1–18.

Ashton, P. (2014). *On the Forests of Tropical Asia: Lest the Memory Fade*. London, Royal Botanic Gardens, Kew.

Bachman, S. P., P. Wilkin, et al. (2020). "Extinction risk and conservation gaps for Aloe (Asphodelaceae) in the Horn of Africa." *Biodiversity and Conservation* 29: 77–98.

Baker, H. G. and I. Baker (1982). Chemical constituents of nectar in relation to pollination mechanisms and phylogeny. *Biochemical Aspects of Evolutionary Biology*. M. H. Niteki. Chicago, University of Chicago Press: 131–171.

Bakken, B. H. and P. Sabat (2008). "The mechanisms and ecology of water balance in hummingbirds." *Ornitologia Neotropica* 19: 501–509.

Baldwin, M. W., Y. Toda, et al. (2014). "Evolution of sweet taste perception in hummingbirds by transformation of the ancestral umami receptor." *Science* 345: 929–933.

Baptista, L. F. and K. L. Schuchmann (1990). "Song learning in the Anna hummingbird (*Calypte anna*)." *Ethology* 84: 15–26.

Barclay, R. M. R. and T. H. Fleming (2020). Reproduction and life histories. *Phyllostomid Bats: a Unique Mammalian Radiation*. T. H. Fleming, L. M. Davalos and M. A. R. Mello. Chicago, University of Chicago Press: 205–220.

Bartley, G. and A. Swash (2022). *Hummingbirds: a Celebration of Nature's Jewels.* Princeton, Princeton University Press.

Bateson, M., S. D. Healy, et al. (2003). "Context-dependent foraging decisions in rufous hummingbirds." *Proceedings of the Royal Society B-Biological Sciences* 270: 1271–1276.

Baum, K. R. and W. E. Grant (2001). "Hummingbird foraging behavior in different patch types: simulations of alternative strategies." *Ecological Modelling* 137: 201–209.

Bell, G. P. and M. B. Fenton (1986). "Visual acuity, sensitivity and binocularity in a gleaning insectivorous bat, *Macrotus californicus* (Chiroptera: Phyllostomidae)." *Animal Behaviour* 34: 409–414.

Berns, C. M. and D. C. Adams (2013). "Becoming different but staying alike: patterns of sexual size and shape dimorphism in bills of hummingbirds." *Evolutionary Biology* 40(2): 246–260.

Beuchat, C. A., W. A. Calder, et al. (1990). "The integration of osmoregulation and energy balance in hummingbirds." *Physiological Zoology* 63: 1059–1081.

Beuchat, C. A. and C. R. Chong (1998). "Hyperglycemia in hummingbirds and its consequences for hemoglobin glycation." *Comparative Biochemistry and Physiology A* 120: 409–420.

Bezemer, N., S. D. Hopper, et al. (2019a). "Primary pollinator exclusion has divergent consequences for pollen dispersal and mating in different populations of a bird-pollinated tree." *Molecular Ecology* 28: 4883–4898.

Bezemer, N., S. L. Krauss, et al. (2019b). "Conservation of old individual trees and small populations is integral to maintain species' genetic diversity of a historically fragmented woody perennial." *Molecular Ecology* 28: 3339–3357.

Bininda-Emonds, O. R. P., M. Cardillo, et al. (2007). "The delayed rise of present-day mammals." *Nature* 446: 507–512.

Bishop, K. L. (2008). "The evolution of flight in bats: narrowing the field of plausible hypotheses." *Quarterly Review of Biology* 83: 153–169.

Blankenship, R. E. (2010). "Early evolution of photosynthesis." *Plant Physiology* 154: 434–438.

Bleiweiss, R. (1998). "Origin of hummingbird faunas." *Biological Journal of the Linnean Society* 65: 77–97.

Bleiweiss, R. (1999). "Joint effects of feeding and breeding behaviour on trophic dimorphism in hummingbirds." *Proceedings of the Royal Society of London Series B–Biological Sciences* 266(1437): 2491–2497.

Bleiweiss, R. (2009). "The tail end of hummingbird evolution: parallel flight system development in living and ancient birds." *Biological Journal of the Linnean Society* 97: 467–493.

Bolzan, D. P., L. M. Pessoa, et al. (2015). "Allometric patterns and evolution in Neotropical nectar-feeding bats (Chiroptera, Phyllostomidae)." *Acta Chiropterologica* 17: 59–73.

Bonaccorso, F. J. and B. K. McNab (1997). "Plasticity of energetics in blossom bats (Pteropodidae): Impact on distribution." *Journal of Mammalogy* 78(4): 1073–1088.

Bonaccorso, F. J., A. Arends, et al. (1992). "Thermal ecology of moustached and mormoopid bats (Mormoopidae) in Venezuela." *Journal of Mammalogy* 73: 365–378.

Bonaccorso, F. J., J. R. Winkelmann, et al. (2002). "Home range of *Dobsonia minor* (Pteropodidae): a solitary, foliage-roosting fruit bat in Papua New Guinea." *Biotropica* (34): 127–135.

Boyle, W. A. and C. J. Conway (2007). "Why migrate? A test of the evolutionary precursor hypothesis." *American Naturalist* 169: 344–359.

Brice, A. T., K. H. Dahl, et al. (1989). "Pollen digestibility by hummingbirds and psittacines." *Condor* 91: 681–688.

Brokaw, A. F., E. Davis, et al. (2021). "Flying bats use serial sampling to locate odour sources." *Biological Letters* 17: 1–6.

Brokaw, A. F. and M. Smotherman (2021). "Olfactory tracking strategies in a neotropical fruit bat." *Journal of Experimental Biology* 224: 1–13.

Brusatte, S. L., J. K. O'Connor, et al. (2015). "The origin and diversification of birds." *Current Biology* 25(19): R888–R898.

Buchmann, S. L. (2015). *The Reason for Flowers: Their History, Culture, Biology, and How They Change Our Lives*. New York, Scribner.

Buchmann, S. L. and G. P. Nabhan (1996). *The Forgotten Pollinators*. Washington, D.C., Island Press.

Bumrungsri, S., D. Lang, et al. (2013). "The dawn bat, *Eonycteris spelea* Dobson (Chiroptera: Pteropodidae) feeds mainly on pollen of economically important food plants in Thailand." *Acta Chiropterologica* 15: 95–104.

Calder, W. A., III (1984). *Size, Function, and Life History*. Cambridge, MA, Harvard University Press.

Campillo, L. C., C. H. Oliveros, et al. (2018). "Genomic data resolve gene tree discordance in spiderhunters (Nectariniidae, *Arachnothera*)." *Molecular Phylogenetics and Evolution* 120: 151–157.

CaraDona, P. J., A. M. Iler, et al. (2014). "Shifts in flowering phenology reshape a subalpine plant community." *Proceedings of the National Academy of Sciences of the United States of America* 111: 4916–4921.

Cardona, T. (2019). "Thinking twice about the evolution of photosynthesis." *Open Biology* 9.

Carpenter, R. E. (1969). "Structure and function of the kidney and the water balance of desert bats." *Physiological Zoology* 42: 288–302.

Carter, G. G., J. M. Ratcliffe, et al. (2010). "Flower bats (*Glossophaga soricina*) and fruit bats (*Carollia perspecillata*) rely on spatial cues over shapes and scents when relocating food." *Plos One* 5: 1–6.

Castro, D. L. J., R. M. P. Freitas, et al. (2021). "Insulin and glucose regulation at rest and during flight in a Neotropical nectar-feeding bat." *Mammalian Biology*. https://doi.org/10.1007/s42991-021-00146-x.

Ceballos, G. and J. H. Brown (1995). "Global patterns of mammalian diversity, endemism, and endangerment." *Conservation Biology* 9(3): 559–568.

Chai, P. and R. Dudley (1996). "Limits to flight energetics of hummingbirds hovering in hypodense and hypoxic gas mixtures." *Journal of Experimental Biology* 199: 2285–2295.

Cheke, R. A., C. F. Mann, and R. Allen. (2001). *Sunbirds, a Guide to the Sunbirds, Flowerpeckers, Spiderhunters, and Sugarbirds of the World*. New Haven, CT, Yale University Press.

Chen, A., N. D. White, et al. (2019). "Total evidence framework reveals complex morphological evolution in nightbirds (Strisores)." *Diversity and Distributions* 9.

Cirranello, A. L. and N. B. Simmons (2020). Diversity and discovery. *Phyllostomid Bats: a Unique Mammalian Radiation*. T. H. Fleming, L. M. Davalos and M. A. R. Mello. Chicago, University of Chicago Press: 43–61.

Cirranello, A. L. and N. B. Simmons (2020). Diversity and discovery: a golden age. *Phyllostomid Bats: a Unique Mammalian Radiation.* T. H. Fleming, L. M. Davalos and M. A. R. Mello. Chicago, University of Chicago Press: 43–62.

Claramunt, S. and J. Cracraft (2015). "A new time tree reveals Earth history's imprint on the evolution of modern birds." *Science Advances* 1(11).

Clark, C. J. (2010). "The evolution of tail shape in hummingbirds." *Auk* 127: 44–56.

Clark, C. J. and R. Dudley (2009). "Flight costs of long, sexually selected tails in hummingbirds." *Proceedings of the Royal Society B-Biological Sciences* 276: 2109–2015.

Clark, C. J., D. O. Elias, et al. (2011). "Aeroelastic flutter produces hummingbird feather songs." *Science* 333: 1430–1433.

Clark, C. J., J. A. McGuire, et al. (2018). "Complex coevolution of wing, tail, and vocal sounds of courting male bee hummingbirds." *Evolution* 72: 630–646.

Cockburn, G., M.-C. Ko, et al. (2014). "Synergism, bifunctionality, and the evolution of a gradual sensory trade-off in hummingbird taste receptors." *Molecular Biology Evolution* 39: 1–12.

Collins, C. T. (2015). "Food habits and resource partitioning in a guild of neotropical swifts." *Wilson Journal of Ornithology* 127: 239–248.

Cox, P. A. (1983). "Extinction of the Hawaiian avifauna resulted in a change of pollinators of the eiei, *Freycinetia arborea*." *Oikos* 41: 195–199.

Crome, F. H. J. and A. K. Irvine (1986). "'Two bob each way': the pollination and breeding system of the Australian rain forest tree *Syzygium cormiflorum*." *Biotropica* 18: 115–125.

Cruz-Neto, A. P. and G. Herrera (2020). The relationship between physiology and diet. *Phyllostomid Bats: a Unique Mammalian Radiation.* T. H. Fleming, L. M. Davalos, and M. A. R. Mello. Chicago, University of Chicago Press: 169–186.

D'Alba, L., M. Meadows, et al. (2021). "Morphogenesis of iridescent feathers in Anna's hummingbird *Calypte anna*." *Integrative and Comparative Biology* 61: 1502–1510.

Dalsgaard, B., P. K. Maruyama, et al. (2021). "The influence of biogeographical and evolutionary histories on morphological trait-matching and resource specialization in mutualistic hummingbird-plant networks." *Functional Ecology* 35: 1120–1133.

Datzmann, T., O. Von Helversen, et al. (2010). "Evolution of nectarivory in phyllostomid bats (Phyllostomidae Gray, 1825, Chiroptera: Mammalia)." *Bmc Evolutionary Biology* 10(165).

Davalos, L. M., A. L. Cirranello, et al. (2020). Adapt or live: adaptation, convergent evolution, and plesiomorphy. *Phyllostomid Bats: a Unique Mammalian Radiation.* T. H. Fleming, L. M. Davalos and M. A. R. Mello. Chicago, University of Chicago Press: 105–121.

Davalos, L. M., P. M. Velazco, et al. (2020). Phylogenetics and historical biogeography. *Phyllostomid Bats: a Unique Mammalian Radiation.* T. H. Fleming, L. M. Davalos and M. A. R. Mello. Chicago, University of Chicago Press: 87–104.

Davies, T. J., S. A. Fritz, et al. (2008). "Phylogenetic trees and the future of mammalian biodiversity." *Proceedings of the National Academy of Sciences of the United States of America* 105: 11556–11563.

Dellinger, A. S. (2020). "Pollination syndromes in the 21st century: where do we stand and where may we go?" *New Phytologist* 228: 1193–1213.

Denzinger, A., M. Tschapka, et al. (2018). "The role of echolocation strategies for niche differentiation in bats." *Canadian Journal of Zoology* 96: 171–181.

Diamant, E. S., J. J. Falk, et al. (2021). "Male-like female morphs in hummingbirds: the evolution of a wide-spread sex-limited plumage polymorphism." *Proceedings of the Royal Society B-Biological Sciences* 288.

Doucette, J. L., R. M. Brigham, et al. (2012). "Prey availability affects daily torpor by free-ranging Australian owlet-nightjars (*Aegotheles cristatus*)." *Oecologia* 169: 361–372.

Driskell, A. C. and L. Christidis (2004). "Phylogeny and evolution of the Australo-Papuan honeyeaters (Passeriformes, Meliphagidae)." *Molecular Phylogenetics and Evolution* 31: 943–960.

Dumont, E. R. (2004). "Patterns of diversity in cranial shape in plant-visiting bats." *Acta Chiropterologica* 6: 59–74.

Dunphy, B. K., J. L. Hamrick, et al. (2004). "A comparison of direct and indirect measures of gene flow in the bat-pollinated tree *Hymenaea courbaril* in the dry forest life zone of southwestern Puerto Rico." *International Journal of Plant Sciences* 165(3): 427–436.

Duque, F. G. and L. L. Carruth (2022). "Vocal communication in hummingbirds." *Brain Behavior and Evolution* 97: 241–252.

Duque, F. G., C. A. Rodríguez-Saltos, et al. (2020). "High-frequency in a hummingbird." *Science Advances* 6: 1–7.

Eberts, E. R., C. G. Guglielmo, et al. (2021). "Reversal of the adipostat control of torpor during migration in hummingbirds." *eLife* 2021: 1–21.

Eby, P. (1991). "Seasonal movements of grey-headed flying foxes, *Pteropus poliocephalus*, from two maternity camps in northern New South Wales." *Wildlife Research* 18: 547–559.

Endress, P. K. (1994). *Diversity and Evolutionary Biology of Tropical Flowers*. Cambridge, UK, Cambridge University Press.

Endress, P. K. (2011). "Evolutionary diversification of the flowers of angiosperms." *American Journal of Botany* 98: 370–396.

Falk, J. J., M. S. Webster, et al. (2021). "Male-like ornamentation in female hummingbirds results from social harassment rather than sexual selection." *Current Biology* 31: 1–7.

Feinsinger, P. and R. K. Colwell (1978). "Community organization among neotropical nectar-feeding birds." *American Zoologist* 18(4): 779–795.

Feldhammer, G. A., J. F. Merritt, et al. (2020). *Mammalogy: Adaptation, Diversity, Ecology*. Baltimore, Johns Hopkins University Press.

Fenton, M. B. (2013). "Questions, ideas and tools: lessons from bat echolocation." *Animal Behaviour* 85: 869–879.

Fleming, P. A., D. A. Gray, et al. (2004). "Osmoregulatory response to acute diet change in an avian nectarivore: rapid rehydration following water shortage." *Comparative Biochemistry and Physiology A—Molecular & Integrative Physiology* 138(3): 321–326.

Fleming, P. A. and S. W. Nicolson (2003). "Osmoregulation in an avian nectarivore, the white-bellied sunbird *Nectarinia talatala*: response to extremes of diet concentration." *Journal of Experimental Biology* 206(11): 1845–1854.

Fleming, T. H. (2006). "Reproductive consequences of early flowering in organ pipe cactus, *Stenocereus thurberi*." *International Journal of Plant Sciences* 167: 473–481.

Fleming, T. H. (2022). "Good to the last drop: feeding behavior of the nectar bat *Leptonycteris yerbabuenae* (Chiroptera, Phyllostomidae) at hummingbird feeders in Tucson, Arizona." *Acta Chiropterologica* 24: 353–361.

Fleming, T. H. (2022). *Sonoran Desert Journeys: Ecology and Evolution of Its Iconic Species*. Tucson, University of Arizona Press.

Fleming, T. H. and W. J. Kress (2013). *The Ornaments of Life, Coevolution and Conservation in the Tropics*. Chicago, University of Chicago Press.

Fleming, T. H. and A. Martino (2020). Population biology. *Phyllostomid Bats: a Unique Mammalian Radiation*. T. H. Fleming, L. M. Davalos and M. A. R. Mello. Chicago, University of Chicago Press: 325–346.

Fleming, T. H. and N. Muchhala (2008). "Nectar-feeding bird and bat niches in two worlds: pantropical comparisons of vertebrate pollination systems." *Journal of Biogeography* 35(5): 764–780.

Fleming, T. H., N. Muchhala, et al. (2005). New World nectar-feeding vertebrates: community patterns and processes. *Contribuciones Mastozoologicas en Homenaje a Bernardo Villa*. V. Sanchez-Cordero and R. A. Medellin. Mexico, Instituto de Biologia y Instituto de Ecologia, Universidad Nactional Autonoma de Mexico: 161–184.

Fleming, T. H. and J. Nassar (2002). Population biology of the lesser long-nosed bat, *Leptonycteris curasoae*, in Mexico and northern South America. *Columnar Cacti and Their Mutualists: Evolution, Ecology, and Conservation*. T. H. Fleming and A. Valiente-Banuet. Tucson, University of Arizona Press: 283–305.

Fleming, T. H., A. A. Nelson, et al. (1998). "Roosting behavior of the lesser long-nosed bat, *Leptonycteris curasoae*." *Journal of Mammalogy* 79(1): 147–155.

Fleming, T. H., C. K. Geiselman, and W. J. Kress (2009). "The evolution of bat pollination: a phylogenetic perspective." *Annals of Botany* 104: 1017–1043.

Fleming, T. H., C. T. Sahley, et al. (2001). "Sonoran Desert columnar cacti and the evolution of generalized pollination systems." *Ecological Monographs* 71: 511–530.

Ford, H. A. and D. C. Paton (1977). "The comparative ecology of ten species of honeyeaters in South Australia." *Australian Journal of Ecology* 2: 399–407.

Forrest, J., D. W. Inouye, et al. (2010). "Flowering phenology in subalpine meadows: does climate variation influence community co-flowering patterns?" *Ecology* 91: 431–440.

Franklin, D. C. and R. A. Noske (1999). "Birds and nectar in a monsoonal woodland: correlations at three spatio-temporal scales." *Emu* 99: 15–28.

Freeman, P. W. (1995). "Nectarivorous feeding mechanisms in bats." *Biological Journal of the Linnean Society* 56: 439–463.

Freeman, P. W. (2000). "Macroevolution in Microchiroptera: Recoupling morphology and ecology with phylogeny." *Evolutionary Ecology Research* 2(3): 317–335.

Freeman, S., L. Allison, et al. (2014). *Biological Science*. Boston, Pearson.

Funamoto, D. (2019). "Plant-pollination interactions in East Asia: a review." *Journal of Pollination Ecology* 25: 46–68.

Futuyma, D. J. and M. Slatkin, eds. (1983). *Coevolution*. Sunderland, MA, Sinauer Associates.

Gahr, M. (2000). "Neural song control system of hummingbirds: comparison to swifts, vocal learning (songbirds), and nonlearning (Suboscines) passerines, and vocal learning (budgerigars) and nonlearning (dove, owl, gull, quail, chicken) nonpasserines." *Journal of Comparative Neurology* 426: 182–196.

Gamba, D. and N. Muchhala (2021). "Pollinator type strongly impacts gene flow within and among plant populations for six neotropical species." *Ecology* 2022: 1–12.

Gardiner, J. D., J. R. Codd, et al. (2011). "An association between ear and tail morphology in bats and their foraging style." *Canadian Journal of Zoology* 89: 90–99.

Gaunt, S. L. L., L. F. Baptista, et al. (1994). "Song learning as evidenced from song sharing in two hummingbird species (*Colibri coruscans* and *C. thalassinus*)." *Auk* 111: 87–103.

Geerts, S., A. Coetzee, et al. (2020). "Pollination structure [of] plant and nectar-feeding bird communities in Cape fynbos, South Africa: implications for the conservation of bird-plant mutualisms." *Ecological Research* 2020: 1–19.

Giannini, N. P. and P. M. Velazco (2020). Phylogeny, fossils, and biogeography: the evolutionary history of Superfamily Noctilionoidea (Chiroptera: Yangochiroptera). *Phyllostomid Bats: a Unique Mammalian Radiation.* T. H. Fleming, L. M. Davalos and M. A. R. Mello. Chicago, University of Chicago Press: 25–42.

Gilbert, L. E. (1980). Food web organization and the conservation of Neotropical diversity. *Conservation Biology.* M. E. Soule and B. A. Wilcox. Sunderland, Massachusetts, Sinauer: 11–33.

Gill, F. B. (1998). "Trapline foraging by hermit hummingbirds: competition for an undefended, renewable resource." *Ecology* 69: 1933–1942.

Gill, F. B. and R. O. Prum (2019). *Ornithology.* New York, W. H. Freeman.

Givnish, T. J., M. H. J. Barfuss, et al. (2014). "Adaptive radiation, correlated and contingent evolution, and net species diversification in Bromeliaceae." *Molecular Phylogenetics and Evolution* 71: 55–78.

Goller, B. and D. L. Altshuler (2014). "Hummingbirds control hovering flight by stabilizing visual motion." *Proceedings of the National Academy of Sciences of the United States of America* 111: 18375–18380.

Gonzalez, C. and J. F. Ornelas (2014). "Acoustic divergence with gene flow in a lekking hummingbird with complex songs." *Plos One* 9: 1–13.

González-Gómez, P. L. and M. Araya-Salas (2019). Perspectives on the study of field hummingbird cognition in the neotropics. *Behavioral Ecology of Neotropical Birds.* J. C. Reboreda, V. D. Fiorini and D. T. Tuero, Springer Nature Switzerland: 199–212.

González-Gómez, P. L., R. A. Vasquez, et al. (2011). "Flexibility of foraging behavior in hummingbirds: the role of energy constraints and cognitive ability." *Auk* 128: 36–42.

Gonzalez-Terrazas, T. P., J. C. Koblitz, et al. (2016a). "How nectar-feeding bats localize their food: echolocation behavior of *Leptonycteris yerbabuenae* approaching cactus flowers." *Plos One* 11(9).

Gonzalez-Terrazas, T. P., C. Martel, et al. (2016b). "Finding flowers in the dark: nectar-feeding bats integrate olfaction and echolocation while foraging for nectar." *Royal Society Open Science* 3(8).

Gorresen, P. M., P. M. Cryan, et al. (2015). "Ultraviolet vision may be widespread in bats." *Acta Chiropterologica* 17: 193–198.

Grace, O. M., R. R. Klopper, et al. (2013). "A revised generic classification for *Aloe* (Xanthorrhoeacea subfam. Asphodeloideae)." *Phytotaxa* 76: 7–14.

Graham, C. H., J. L. Parra, et al. (2009). "Phylogenetic structure in tropical hummingbird communities." *Proceedings of the National Academy of Sciences of the United States of America* 106: 19673–19678.

Greenewalt, C. H. (1960). *Hummingbirds.* Garden City, Doubleday and the American Museum of Natural History.

Groom, D. J. E., M. C. B. Toledo, et al. (2017). "Wing kinematics and energetics during weight-lifting in hovering in hummingbirds across an elevational gradient." *Journal of Comparative Physiology B.*

Hall, E. R. and K. R. Kelson (1959). *The Mammals of North America, Vol. 1.* New York, The Ronald Press Company.

Hall, L. S. and G. Richards (2000). *Flying Foxes: Fruit and Blossom Bats of Australia.* Malabar, FL, Krieger Publishing Company.

Hall, R. P., G. L. Mutumi, et al. (2021). "Find the food first: an omnivorous sensory morphotype predates biomechanical specialization for plant based diets in phyllostomid bats." *Evolution* 2021: 1–11.

Hargrove, J. L. (2005). "Adipose energy stores, physical work, and the metabolic syndrome: lessons from hummingbirds." *Nutrition Journal* 2005.

Hassanin, A., C. Bonillo, et al. (2020). "Phylogeny of African fruit bats (Chiroptera, Pteropodidae) based on complete mitochondrial genomes." *Journal of Zoological Systematics and Evolutionary Research* 58: 1395–1410.

Heithaus, E. R., T. H. Fleming, et al. (1975). "Patterns of foraging and resource utilization in seven species of bats in a seasonal tropical forest." *Ecology* 56: 841–854.

Henry, M. and K. E. Stoner (2011). "Relation between spatial working memory performance and diet specialization in two sympatric nectar bats" *Plos One* 6: 1–7.

Hernández-Hernández, T., J. W. Brown, et al. (2014). "Beyond aridification: multiple explanations for the elevated diversification of cacti in the New World Succulent Biome." *New Phytologist* 202: 1382–1397.

Herrera, G., J. C. Zagal, et al. (2008). "Spectral sensitivities of photoreceptors and their role in color discrimination in the green-backed firecrown hummingbird (*Sephanoides sephanoides*)." *Journal of Comparative Physiology A* 194: 785–794.

Hileman, L. C. (2014). "Trends in flower symmetry evolution revealed through phylogenetic and developmental genetic advances." *Philosophical Transactions of the Royal Society B* 369: 1–10.

Hixon, M. A., F. L. Carpenter, et al. (1983). "Territory area, flower density, and time budgeting in hummingbirds: an experimental and theoretical analysis." *American Naturalist* 122: 366–391.

Hodgkison, R., S. T. Balding, et al. (2003). "Fruit bats (Chiroptera) as seed dispersers and pollinators in a lowland Malaysian rain forest." *Biotropica* 35: 491–502.

Hopper, S. D., H. Lambers, et al. (2021). "OCBIL theory examined: reassessing evolution, ecology, and conservation in the world's ancient, climatically buffered and infertile landscapes." *Biological Journal of the Linnean Society* 133: 266–296.

Hopper, S. D. and G. F. Moran (1981). "Bird pollination and the mating system of *Eucalyptus stoatei*." *Australian Journal of Botany* 29: 625–638.

Horner, M. A., T. H. Fleming, et al. (1998). "Foraging behaviour and energetics of a nectar-feeding bat, *Leptonycteris curasoae* (Chiroptera: Phyllostomidae)." *Journal of Zoology* 244: 575–586.

Hughes, M., M. Moller, et al. (2007). "The impact of pollination syndrome and habitat on gene flow: a comparative study of two *Streptocarpus* (Gesneriaceae) species." *American Journal of Botany* 94: 1688–1695.

Ibbotson, M. R. (2017). "Visual neuroscience: unique neural system for flight stabilization in hummingbirds." *Current Biology* 27: R57–R76.

Ingersoll, R., L. Haizmann, et al. (2018). "Biomechanics of hover performance in neotropical hummingbirds versus bats." *Science Advances* 4.

Ioale, P. and F. Papi (1989). "Olfactory bulb size, odor discrimination, and magnetic insensitivity in hummingbirds." *Physiology and Behavior* 45: 995–999.

Irish, V. (2017). "The ABC model of floral development." *Current Biology* 27: R853–R909.

Jaggard, A. K., N. Smith, et al. (2015). "Rules of the roost: characteristics of nocturnal communal roosts of rainbow lorikeets (*Trichoglossus haematodus*, Psitaccidae) in an urban environment." *Urban Ecosystems* 2015: 1–14.

Janacek, S., K. Chmel, et al. (2021). "Spatio-temporal pattern of specialization of sunbird-flower networks on Mt. Cameroon." *Research Square*: 1–19.

Janzen, D. H. (1980). "When is it coevolution?" *Evolution* 34: 611–612.

Jarvis, E. D. (2004). "Learned birdsong and the neurobiology of human language." *Annals of the New York Academy of Sciences* 1016: 749–777.

Jarvis, E. D. and A. B. N. Consortium (2005). "Avian brains and a new understanding of vertebrate brain evolution." *Nature Reviews* 6: 151–159.

Jarvis, E. D., S. Mirarab, et al. (2014). "Whole-genome analyses resolve early branches in the tree of life of modern birds." *Science* 346: 1320–1331.

Jarvis, E. D., S. Ribiero, et al. (2000). "Behaviorally driven gene expression reveals song nuclei in hummingbird brain." *Nature* 406: 628–632.

Johnsgard, P. A. (2016). *The Hummingbirds of North America*. Washington, Smithsonian Institution Press.

Johnson, K. E. and C. J. Clark (2020). "Ontogeny of vocal learning in a hummingbird." *Animal Behaviour* 167: 139–150.

Jones, G. and E. C. Teeling (2006). "The evolution of echolocation in bats." *Trends in Ecology & Evolution* 21: 149–156.

Jones, K. E., S. P. Mickleburgh, et al. (2009). Global overview of the conservation of island bats: importance, challenges, and opportunities. *Island Bats: Evolution, Ecology, and Conservation*. T. H. Fleming and P. Racey. Chicago, University of Chicago Press: 496–537.

Joseph, L., A. Toon, et al. (2014). "A new synthesis of the molecular systematics and biogeography of honeyeaters (Passeriformes, Meliphagidae) highlights biogeographical and ecological complexity of a spectacular avian radiation." *Zoologica Scripta* 43: 235–248.

Kalko, E. K. V. and H. U. Schnitzler (1998). How echolocating bats approach and acquire food. *Bat Biology and Conservation*. T. H. Kunz and P. Racey. Washington, Smithsonian Institution Press: 197–204.

Karasov, W. H. and S. J. Cork (1996). "Test of a reactor-based digestion optimization model for nectar-eating rainbow lorikeets." *Physiological Zoology* 69: 117–138.

Kelm, D. H. and O. v. Helversen (2007). "How to budget metabolic energy: torpor in a small Neotropical mammal." *Journal of Comparative Biology B* 177: 667–677.

Kim, A. V., D. T. Rankin, et al. (2021). "What is that smell? Hummingbirds avoid foraging on resources with defensive insect compounds." *Behavioral Ecology and Sociobiology* 75: 1–9.

Knoll, A. H. (2021). *A Brief History of Earth*. New York, Mariner Books.

Knudsen, J. T., L. Tollsten, et al. (2004). "Trends in floral scent chemistry in pollination syndromes: floral scent composition in hummingbird-pollinated taxa." *Botanical Journal of the Linnean Society* 46: 191–199.

Kodric-Brown, A., J. H. Brown, et al. (1984). "Organization of a tropical island community of hummingbirds and flowers." *Ecology* 65: 1358–1368.

Kries, K., M. A. S. Barros, et al. (2018). "Colour vision variation in leaf-nosed bats (Phyllostomidae): links to cave roosting and dietary specialization." *Molecular Ecology* 27: 3627–3640.

Kruyt, J. W., E. M. Quicazano-Rubio, et al. (2014). "Hummingbird wing efficacy depends on aspect ratio and compares with helicopter rotors." *Journal of the Royal Society Interface* 11.

Ksepka, D. T., J. A. Clarke, et al. (2013). "Fossil evidence of wing shape in a stem relative of swifts and hummingbirds (Aves, Pan-Apodiformes)." *Proceedings of the Royal Society B-Biological Sciences* 280(1761).

Lagomarsino, L. P., F. L. Condamine, et al. (2016). "The abiotic and biotic drivers of rapid diversification in Andean bellflowers (Campanulaceae)." *New Phytologist* 210(4): 1430–1442.

Lagomarsino, L. P., E. J. Forrestel, et al. (2017). "Repeated evolution of vertebrate pollination syndromes in a recently diverged Andean plant clade." *Evolution* 71(8): 1970–1985.

Lara, C. (2006). "Temporal dynamics of flower use by hummingbirds in a highland temperate forest in Mexico." *Ecoscience* 13: 23–29.

Lauron, E. J., C. Loiseau, et al. (2014). "Coevolutionary patterns and diversification of avian malaria parasites in African sunbirds (Family Nectariniidae)." *Parasitology International* 2014: 1–13.

Law, B. S. (1993). "Roosting and foraging ecology of the Queensland blossom bat (*Syconycteris australis*) in north-eastern New-South Wales—flexibility in response to seasonal variation." *Wildlife Research* 20(4): 419–431.

Law, B. S. and M. Lean (1999). "Common blossom bats (*Syconycteris australis*) as pollinators in fragmented Australian tropical rainforest." *Biological Conservation* 91(2–3): 201–212.

Lee, D. (2007). *Nature's Palette, the Science of Plant Color*. Chicago, University of Chicago Press.

Leimberger, K. G., B. Dalsgaard, et al. (2022). "The evolution, ecology, and conservation of hummingbirds and their interactions with flowering plants." *Biological Reviews* 97: 923–959.

Lemke, T. O. (1984). "Foraging ecology of the long-nosed bat, *Glossophaga soricina*, with respect to resource availability." *Ecology* 65: 538–548.

Leopold, A. (1972). *Round River*. New York, Oxford University Press.

Levey, D. J. and F. G. Stiles (1992). "Evolutionary precursors of long-distance migration: resource availability and movement patterns in neotropical landbirds." *American Naturalist* 140: 447–476.

Lim, M. C. W., K. Bi, et al. (2019). "Pervasive genomic signatures of local adaptation to altitude across highland specialist Andean hummingbird populations." *Journal of Heredity* 112: 229–240.

Lopes, S. A., P. J. Bergamo, et al. (2022). "Heterospecific pollen deposition is positively associated with reproductive success in a diverse hummingbird-pollinated plant community." *Oikos* 2022: e08714.

Lotz, C. N. and C. Martínez del Rio (2004). "The ability of rufous hummingbirds *Selasphorus rufus* to dilute and concentrate urine." *Journal of Avian Biology* 35: 54–62.

Lovette, I. J., E. Bermingham, et al. (2002). "Clade-specific morphological diversification and adaptive radiation in Hawaiian songbirds." *Proceedings of the Royal Society of London Series B-Biological Sciences* 269(1486): 37–42.

Low, T. (2016). *Where Song Began: Australia's Birds and How They Changed the World*. New Haven, Yale University Press.

Low, Y. W., S. Rajaraman, et al. (2022). "Genomic insights into rapid speciation within the world's largest tree genus *Syzgium*." *Nature Communications* 13: 1–15.

Maeda, M., T. Nikata, et al. (2017). "Quantifying the dynamic wing morphing of hovering hummingbird." *Royal Society Open Science* 4.

Magallón, S., S. Gomez-Acevedo, et al. (2015). "A metacalibrated time-tree documents the early rise of flowering plant phylogenetic diversity." *New Phytologist* 207: 437–453.

Manning, J., J. S. Boatwright, et al. (2014). "A molecular and generic classification of Asphodelaceae subfamily Alooideae: a final resolution of the prickly issue of polyphyly in the Alooids." *Systematic Botany* 39: 55–74.

Markel, H. (2021). *The Secret of Life: Rosalind Franklin, James Watson, Francis Crick, and the Discovery of DNA's Double Helix*. New York, W. W. Norton & Company, Inc.

Marki, P. Z., J. D. Kennedy, et al. (2019). "Adaptive radiation and the evolution of nectarivory in a large songbird clade." *Evolution* 73: 1226–1240.

Martén-Rodríguez, S., A. Almares-Castro, et al. (2009). "Evaluation of pollination syndromes in Antillean Gesneriaceae: evidence for bat, hummingbird and generalized flowers." *Journal of Ecology* 97: 348–359.

Martén-Rodríguez, S., C. B. Fenster, et al. (2010). "Evolutionary breakdown of pollination specialization in a Caribbean plant radiation." *New Phytologist* 188(2): 403–417.

Martin, W. F., T. A. Bryant, et al. (2018). "A physiology perspective on the origin and evolution of photosynthesis." *FEMS Microbiology Reviews* 42: 205–231.

Mauseth, J. D. (2021). *Botany: an Introduction to Plant Biology*. Burlington, MA, Jones & Bartlett Learning LLC.

Mayr, G. (2003). "A new Eocene swift-like bird with a peculiar feathering." *Ibis* 145: 382–391.

Mayr, G. (2004). "Old World fossil record of modern-type hummingbirds." *Science* 304: 861–864.

Mayr, G. and V. Wilder (2004). "Eocene fossil is earliest evidence of flower-visiting by birds." *Biology Letters* 10. https://doi.org/10.1098/rsbl.2014.0223.

McCracken, G. F. and G. S. Wilkinson (2000). Bat mating systems. *Reproductive Biology of Bats*. E. G. Crichton and P. H. Krutzsch. San Diego, Academic Press: 321–362.

McGuire, J. A., C. C. Witt, et al. (2014). "Molecular phylogenetics and the diversification of hummingbirds." *Current Biology* 24: 910–916.

McKinney, A. M., P. J. CaraDona, et al. (2012). "Asynchronous changes in phenology of migrating broad-tailed hummingbirds and their early season nectar resources." *Ecology* 93: 1987–1993.

McNab, B. K. (1986). "The influence of food habits on the energetics of eutherian mammals." *Ecological Monographs* 56: 1–19.

McNab, B. K. (1988). "Food habits and the basal rate of metabolism in birds." *Oecologia* 77: 343–349.

McNab, B. K. (2002). *The physiological ecology of vertebrates*. Ithaca, New York, Cornell University Press.

McNab, B. K. (2003). "Standard energetics of phyllostomid bats: the inadequacies of phylogenetic-contrast analyses." *Comparative Biochemistry and Physiology A—Molecular & Integrative Physiology* 135A: 357–368.

McNab, B. K. (2012). *Extreme Measures: The Ecological Energetics of Birds and Mammals*. Chicago, University of Chicago Press.

McNab, B. K. and F. J. Bonaccorso (1995). "The energetics of Australian swifts, frogmouths, and nightjars." *Physiological Zoology* 68: 245–261.

McNab, B. K. and F. J. Bonaccorso (2001). "The metabolism of New Guinean pteropodid bats." *Journal of Comparative Physiology A* 171: 201–214.

McWhorter, T. J. and C. Martínez del Rio (1999). "Food ingestion and water turnover in hummingbirds: how much dietary water is absorbed?" *Journal of Experimental Biology* 202: 2851–2858.

Medina-Tapia, N., J. Alaya-Berdon, et al. (2012). "Do hummingbirds have a sweet tooth? Gustatory sugar thresholds and sugar selection in the broad-billed hummingbird *Cynanthis latirostris*." *Comparative Biochemistry and Physiology A* 161: 307–314.

Menz, M. H. M., R. D. Phillips, et al. (2011). "Reconnecting plants and pollinators: challenges in the restoration of pollination mutualisms." *Trends in Plant Science* 16(1): 4–12.

Mitchell, W. A. (1989). "Informational constraints on optimally foraging hummingbirds." *Oikos* 55: 145–154.

Monte, A., M. L. da Silva, et al. (2023). "Absence of song suggests heterogeneity of vocal-production learning in hummingbirds." *Journal of Ornithology*.

Monteiro, L. R. and M. R. Nogueira (2011). "Evolutionary patterns and processes in the radiation of phyllostomid bats." *Bmc Evolutionary Biology* 11: 1–23.

Moran, A. J., S. W. J. Prosser, et al. (2019). "DNA metabarcoding allows non-invasive identification of arthropod prey provisioned to nestling rufous hummingbirds (*Selasphorus rufus*)." *PeerJ* 7.

Moyle, R. G., S. S. Taylor, et al. (2011). "Diversification of an endemic Southeast Asian genus: phylogenetic relationships of the spiderhunters (Nectariniidae: *Arachnothera*)." *Auk* 128(4): 777–788.

Moyroud, E. and B. J. Glover (2017). "The evolution of diverse floral morphologies." *Current Biology* 27: R941–R951.

Muchhala, N. and D. Serrano (2015). "The complexity of background clutter affects nectar bat use of flower odor and shape cues." *Plos One* 10: 1–12.

Muchhala, N. and M. Tschapka (2020). The ecology and evolution of nectar feeders. *Phyllostomid Bats: a Unique Mammalian Radiation*. T. H. Fleming, L. M. Davalos and M. A. R. Mello. Chicago, University of Chicago Press: 273–295.

Munguia-Rosas, M. A., V. J. Sosa, et al. (2009). "Specialization clines in the pollination systems of Agaves (Agavaceae) and columnar cacti (Cactaceae): a phylogenetically controlled meta-analysis." *American Journal of Botany* 96(10): 1887–1895.

Murray, K. L. and T. H. Fleming (2008). "Social structure and mating system of the buffy flower bat, *Erophylla sezekorni* (Chiroptera, Phyllostomidae)." *Journal of Mammalogy* 89(6): 1391–1400.

Mutumi, G. L., R. P. Hall, et al. (2023). "Disentangling mechanical and sensory modules in the radiation of noctilionoid bats." *American Naturalist* 202. https://doi.org/10.1086/725368.

Nagy, K. A., I. A. Girard, et al. (1999). "Energetics of free-ranging mammals, reptiles, and birds." *Annual Review of Nutrition* 19: 247–277.

Nan, Y., Y. Chen, et al. (2020). "Experimental studies of tail shapes for hummingbird-like flapping wing micro air vehicles." *IEEE Access*.

Nassar, J. M., L. F. Aguirre, et al. (2020). Threats, status, and conservation perspectives for leaf-nosed bats. *Phyllostomid Bats: a Unique Mammalian Radiation*. T. H. Fleming, L. M. Davalos and M. A. R. Mello. Chicago, University of Chicago Press: 435–456.

Navalón, G., J. Marugan-Lobo, et al. (2020). "The consequences of craniofacial integration for the adaptive radiations of Darwin's finches and Hawaiian honeycreepers." *Nature Ecology & Evolution* 4.

Navalón, G., S. M. Nebrada, et al. (2021). "Craniofacial development illuminates the evolution of nightbirds (Strisores)." *Proceedings of the Royal Society B-Biological Sciences* 288.

Nicolson, S. W. and P. A. Fleming (2003). "Nectar as food for birds: the physiological consequences of drinking dilute sugar solutions." *Plant Systematics and Evolution* 238(1–4): 139–153.

Nocera, J. J., J. M. Blais, et al. (2012). "Historical pesticide applications coincided with an altered diet of aerially foraging insectivorous chimney swifts." *Proceedings of the Royal Society B—Biological Sciences* 279: 3114–3120.

Norberg, U. M. and J. M. V. Rayner (1987). "Ecological morphology and flight in bats (Mammalia; Chiroptera): wing adaptations, flight performance, foraging strategy, and echolocation." *Philosophical Transactions of the Royal Society of London Series B-Biological Sciences* 316: 335–427.

Nowicki, S. and W. A. Searcy (2014). "The evolution of vocal learning." *Current Opinion in Neurobiology* 28: 48–53.

Nsor, C. A., W. Godsoe, et al. (2019). "Promiscuous pollinators—evidence from an Afromontane sunbird-plant pollen transport network." *Biotropica* 51: 538–548.

Nuñez, P., M. M. Méndez, and I. López-Rull (2021). "Can foraging hummingbirds use smell? A test with the Amazilia hummingbird *Amazilia amazilia*." *Ardeola* 68: 433–444.

Ocampo, D., G. Barrantes, et al. (2018). "Morphological adaptations for relatively larger brains in hummingbird skulls." *Ecology and Evolution* 2018: 8: 10482–10488.

O'Leary, M. A., J. I. Bloch, et al. (2013). "The placental mammal ancestor and the post-K-Pg radiation of placentals." *Science* 339(6120): 662–667.

Oliveros, C. H., D. J. Field, et al. (2019). "Earth history and the passerine superradiation." *Proceedings of the National Academy of Sciences of the United States of America* 116: 7916–7925.

Orlowski, G. and J. Karg (2013). "Diet breadth and overlap in three sympatric aerial insectivorous birds at the same location." *Bird Study* 60: 475–483.

Osipova, E., R. Barsacchi, et al. (2023). "Loss of a gluconeogenic muscle enzyme contributed to adaptive metabolic traits in hummingbirds." *Science* 379: 185–190.

Pauw, A. (2019). "A birds-eye view of pollination: biotic interactions as drivers of adaptation and community change." *Annual Review of Ecology, Evolution, and Systematics* 50: 477–502.

Perlov, D. and A. Vilenkin (2017). *Cosmology for the Curious*. Cham, Switzerland, Springer.

Perrins, C. M. and A. L. A. Middleton, eds. (1985). *The Encyclopedia of Birds*. New York, Facts on File.

Pigot, A. L. and R. S. Etienne (2015). "A new dynamic null model for phylogenetic community structure." *Ecology Letters* 18: 153–163.

Potter, J. H. T., R. Drinkwater, et al. (2021a). "Nectar-feeding birds and bats show parallel molecular adaptations in sugar metabolism enzymes." *Current Biology* 31: 4667–4674.

Potter, J. H. T., K. T. J. Davies, et al. (2021b). "Dietary diversification and specialization in neotropical bats facilitated by early molecular evolution." *Molecular Biology and Evolution* 38: 3864–3883.

Powers, D. R. and K. A. Nagy (1988). "Field metabolic rate and food consumption by free-living Anna's hummingbirds (*Calypte anna*)." *George Fox University Faculty Publications-Department of Biology and Chemistry* (No. 28): 1–7.

Prieto-Torres, D. A., L. E. Nunez Rosas, et al. (2022). "Most Mexican hummingbirds lose under climate and land-use change: long-term conservation implications." *Perspectives in Ecology and Conservation* 19: 487–499.

Proctor, M., P. Yeo, et al. (1996). *The Natural History of Pollination*. Portland, Oregon, Timber Press.

Projecto-Garcia, J., C. Natarajan, et al. (2013). "Repeated elevational transitions in hemoglobin function during the evolution of Andean hummingbirds." *Proceedings of the National Academy of Sciences of the United States of America* 110: 20669–20674.

Provost, K. L., L. Joseph, et al. (2018). "Resolving a phylogenetic hypothesis for parrots: implications from systematics to conservation." *Emu* 2018: 1–16.

Prum, R. O. (2017). *The Evolution of Beauty: How Darwin's Forgotten Theory of Mate Choice Shapes the Animal World—and Us*. New York, Doubleday.

Prum, R. O., J. S. Berv, et al. (2015). "A comprehensive phylogeny of birds (Aves) using targeted next-generation DNA sequencing." *Nature* 526(7574): 569-U247.

Pyke, G. H. (1980). "The foraging behaviour of Australian honeyeaters: a review and some comparisons with hummingbirds." *Australian Journal of Ecology* 5: 343–369.

Pyke, G. H. (1984). "Optimal foraging theory: a critical review." *Annual Review of Ecology and Systematics* 15: 523–575.

Pyke, G. H. (2016). "Plant-pollinator coevolution: it's time to connect optimal foraging theory with evolutionary stable strategies." *Plant Ecology, Evolution, and Systematics* 19: 70–76.

Quammen, D. (2018). *The Tangled Tree: a Radical New History of Life*. New York, Simon & Schuster Paperbacks.

Quesada, M., K. E. Stoner, et al. (2004). "Effects of forest fragmentation on pollinator activity and consequences for plant reproductive success and mating patterns in bat-pollinated bombacaceous trees." *Biotropica* 36(2): 131–138.

Raguso, R. A. (2008). "Wake up and smell the roses: the ecology and evolution of floral scent." *Annual Review of Ecology Evolution and Systematics* 39: 549–570.

Rankin, D. T., C. J. Clark, et al. (2018). "Hummingbirds use taste and touch to discriminate against nectar resources that contain Argentine ants." *Behavioral Ecology and Sociobiology* 72: 1–9.

Rappole, J. H. and K. L. Schuchmann (2003). Ecology and evolution of hummingbird population movements and migration. *Avian Migration*. P. Berthold, E. Gwinner and E. Sonnenschein. Berlin, Springer-Verlag: 39–51.

Ratcliffe, J. M. (2009). "Neuroecology and diet selection in phyllostomid bats." *Behavioural Processes* 80: 247–251.

Ratto, F., B. I. Simmons, et al. (2018). "Global importance of vertebrate pollinators for plant reproductive success: a meta-analysis." *Frontiers in Ecology and the Environment* 16(2): 82–90.

Recher, H. F. and W. E. Davis, Jr. (2011). "Observations on the foraging ecology of honeyeaters (Meliphagidae) at Dryandra Woodland, Western Australia." *Amytormis* 3: 19–29.

Remolina-Figueroa, D., D. A. Prieto-Torres, et al. (2022). "Together forever? Hummingbird-plant relationships in the face of climate warming." *Climate Change* 175: 1–21.

Rhodes, R. (1986). *The Making of the Atomic Bomb*. New York, Simon and Schuster, Inc.

Ricklefs, R. E. (2004). "The cognitive face of avian life histories." *Wilson Bulletin* 116: 119–133.

Rico-Guevara, A. and M. Araya-Salas (2015). "Bills as daggers? A test for sexually dimorphic weapons in a lek-mating hummingbirds." *Behavioral Ecology* 26: 21–29.

Rico-Guevara, A., K. J. Hurme, et al. (2021). "Bene'Fit' assessment in pollination coevolution: mechanistic perspectives in hummingbird bill-flower matching." *Integrative and Comparative Biology*.

Rico-Guevara, A., K. J. Hurme, et al. (2023). "Nectar feeding beyond the tongue: hummingbirds drink using phase-shifted bill opening, flexible tongue flaps and wringing at the tips." *Journal of Experimental Biology* 226.

Rico-Guevara, A., M. A. Rubega, et al. (2017). "Shifting paradigms in the mechanics of nectar extraction and hummingbird bill morphology." *Integrative Organismal Biology*.

Riede, T. and C. R. Olson (2020). "The vocal organ of hummingbirds shows convergence with songbirds." *Scientific Reports* 10: 1–14.

Roalson, E. H. and W. R. Roberts (2016). "Distinct processes drive diversification in different clades of Gesneriaceae." *Systematic Biology* 65(4): 662–684.

Rodríguez-Flores, C. I., J. F. Ornelas, et al. (2019). "Are hummingbirds generalists or specialists? Using network analysis to explore the mechanisms influencing their interaction with nectar resources." *Plos One* 14: 1–32.

Rojas, D., C. A. Mancina, et al. (2013). "Phylogenetic signal, feeding behaviour and brain volume in neotropical bats." *Journal of Evolutionary Biology* 26: 1925–1933.

Ronse de Craene, L. (2018). "Understanding the role of floral development in the evolution of angiosperm flowers: clarifications from a historical and physico-dynamic perspective." *Journal of Plant Research* 131: 367–393.

Rosas-Guerrero, V., R. Aguilar, et al. (2014). "A quantitative review of pollination syndromes: do floral traits predict effective pollinators?" *Ecology Letters* (2014): 1–13.

Roxburgh, L. and B. Pinshow (2000). "Nitrogen requirements of an old world nectarivore, the orange-tufted sunbird *Nectarinia osea*." *Physiological and Biochemical Zoology* 73(5): 638–645.

Rubalcaba, J. G., S. F. Gouveia, et al. (2022). "Physical constraints on thermoregulation and flight drive morphological evolution in bats." *Proceedings of the National Academy of Sciences of the United States of America* 119: 1–7.

Ruf, T. and F. Geiser (2015). "Daily torpor and hibernation in birds and mammals." *Biological Reviews* 90: 891–926.

Safi, K., M. V. Cianciaruso, et al. (2011). "Understanding global patterns of mammalian functional and phylogenetic diversity." *Philosophical Transactions of the Royal Society B Biology* 366: 2536–2544.

Safi, K. and D. K. N. Dechmann (2005). "Adaptation of brain regions to habitat complexity: a comparative analysis in bats (Chiroptera)." *Proceedings of the Royal Society B-Biological Sciences* 272: 179–186.

Salewski, V. and B. Bruderer (2007). "The evolution of bird migration—a synthesis." *Naturwissenschaften* 94: 267–279.

Sampson, J. F., S. D. Hopper, et al. (1989). "The mating system and population genetic structure in a bird-pollinated mallee, *Eucalyptus rhodnantha*." *Heredity* 63: 383–393.

Sampson, J. F., S. D. Hopper, et al. (1995). "The mating system and genetic diversity of the Australian arid zone mallee, *Eucalyptus rameliana*." *Australian Journal of Botany* 43: 461–474.

Sapir, N. and R. Dudley (2012). "Backward flight in hummingbirds employs unique kinematic adjustments and entails low metabolic cost." *Journal of Experimental Biology* 215: 3603–3611.

Sapir, N. and R. Dudley (2013). "Implications of floral orientation for flight kinematics and metabolic expenditure of hover-feeding hummingbirds." *Functional Ecology* 27(1): 227–235.

Schnitzler, H. U. and E. K. V. Kalko (1998). How echolocating bats search for and find food. *Bat Biology and Conservation*. T. H. Kunz and P. Racey. Washington, Smithsonian Institution Press: 183–196.

Schoener, T. W. (1971). "Theory of feeding strategies." *Annual Review of Ecology and Systematics* 2: 369–404.

Schondube, J. E., L. G. Herrera-M, et al. (2001). "Diet and the evolution of digestion and renal function in phyllostomid bats." *Zoology-Analysis of Complex Systems* 104(1): 59–73.

Schrödinger, E. (1944). *What Is Life?* Cambridge, UK, Cambridge University Press.

Schuchmann, K. L. (1999). Family Trochilidae. *Handbook of Birds of the World*. J. del Hoya, A. Elliot and J. Sargatal. Barcelona, Lynx Edicions. 5: 468–535.

Schulenberg, T. S., D. F. Stotz, et al. (2007). *Birds of Peru*. Princeton, NJ, Princeton University Press.

Schweizer, M., T. F. Wright, et al. (2015). "Molecular phylogenetics suggests a New Guinean origin and frequent episodes of founder-event speciation in the nectarivorous lories and lorikeets (Aves: Psittaciformes)." *Molecular Phylogenetics and Evolution* 90: 34–48.

Serrano-Serrano, M. L., J. Rolland, et al. (2017). "Hummingbird pollination and the diversification of angiosperms: an old and successful association in Gesneriaceae." *Proceedings of the Royal Society B-Biological Sciences* 284(1852).

Shankar, A., C. H. Graham, et al. (2019). "Hummingbirds budget energy flexibly in response to changing resources." *Functional Ecology* 33: 1904–1916.

Shen, Y.-Y., L. Liang, et al. (2010). "Adaptive evolution of energy metabolism genes and the origin of flight in bats." *Proceedings of the National Academy of Sciences of the United States of America* 107: 8666–8671.

Shi, J. J. and D. L. Rabosky (2015). "Speciation dynamics during the global radiation of extant bats." *Evolution* 69: 1528–1545.

Shubin, N., C. Tabin, et al. (1997). "Fossils, genes, and the evolution of animal limbs." *Nature* 388: 639–648.

Simmons, N. B., G. F. Gunnell, et al. (2020). Fragments and gaps: the fossil record. *Phyllostomid Bats: a Unique Mammalian Radiation*. T. H. Fleming, L. M. Davalos and M. A. R. Mello. Chicago, University of Chicago Press: 63–86.

Simmons, N. B., K. L. Seymour, et al. (2008). "Primitive early Eocene bat from Wyoming and the evolution of flight and echolocation." *Nature* 451: 818–822.

Simon, R., F. Matt, et al. (2023). "An ultrasound absorbing inflorescence zone enhances echo-acoustic contrast of bat-pollinated cactus flowers." *Journal of Experimental Biology* 226: jeb 245263.

Skandalis, D. A., P. S. Segre, et al. (2017). "The biomechanical origin of extreme wing allometry in hummingbirds." *Nature Communications* 8.

Smith, B. T., W. M. Mauck, III, et al. (2020). "Uneven missing data skew phylogenomic relationships within the lories and lorikeets." *Genome Biological Evolution* 12: 1131–1147.

Smith, J. and A. Lill (2008). "Importance of eucalypts in exploitation of urban parks by rainbow and musk lorikeets." *Emu* 108: 187–195.

Sol, D., N. Garcia, et al. (2010). "Evolutionary divergence in brain size between migratory and resident birds." *Plos One* 5: 1–8.

Soltis, D. E., C. D. Bell, et al. (2008). "Origin and early evolution of angiosperms." *Annals of the New York Academy of Sciences* 1133: 3–25.

Soltis, D. E., P. S. Soltis, et al. (2018). *Phylogeny and Evolution of the Angiosperms*. Chicago, University of Chicago Press.

Sonne, J., A. M. M. Gonzalez, et al. (2016). "High proportion of smaller ranged hummingbird species coincides with ecological specialization across the Americas." *Proceedings of the Royal Society B-Biological Sciences* 283(1824): 20152512.

Spence, A. R. and M. W. Tingley (2021). "Body size and environment influence both intraspecific and interspecific variation in daily torpor use in hummingbirds." *Functional Ecology* 2021.

Stanford, L. and A. Lill (2008). "Out on the town: winter feeding of lorikeets in urban parkland." *Corella* 32: 49–57.

Start, A. N. and A. G. Marshall (1976). Nectarivorous bats as pollinators of trees in West Malaysia. *Tropical trees, variation, breeding, and conservation*. J. Burley and B. T. Styles. London, Academic Press: 141–150.

Stewart, A. B. and M. R. Dudash (2018). "Foraging strategies of generalist and specialist Old World nectar bats in response to temporally variable floral resources." *Biotropica* 50: 98–105.

Stiles, F. G. (1978). "Ecological and evolutionary implications of bird pollination." *American Zoologist* 18: 715–727.

Stiles, F. G. (1980). "The annual cycle in a tropical wet forest hummingbird community." *Ibis* 122: 322–343.

Stiles, F. G. (1995). "Behavioral, ecological, and morphological correlates for foraging for arthropods by the hummingbirds of a tropical wet forest." *Condor* 97: 853–878.

Stiles, F. G. (2004). "Phylogenetic constraints upon morphological and ecological adaptation in hummingbirds (Trochilidae): why are there no hermits in the paramo?" *Ornitologia Neotropical* 15: 191–198.

Stiles, F. G. (2008). "Ecomorphology and phylogeny of hummingbirds: divergence and convergence in adaptations to high elevations." *Ornitologia Neotropical* 19: 511–519.

Stiles, F. G. and L. L. Wolf (1979). "Ecology and evolution of lek mating behavior in the long-tailed hermit hummingbird." *Ornithological Monographs* 27: 1–78.

Suarez, R. K. (1992). "Hummingbird flight: sustaining the highest mass-specific metabolic rates in vertebrates." *Experientia* 98: 565–570.

Suarez, R. K. (1998). "Oxygen and the upper limits to animal design and performance." *Journal of Experimental Biology* 201: 1065–1072.

Suarez, R. K. (2013). "Premigratory fat metabolism in hummingbirds: a Rumsfeldian approach." *Current Biology* 59: 371–380.

Suarez, R. K. and K. C. Welch (2017). "Sugar metabolism in hummingbirds and nectar bats." *Nutrients* 9(7).

Suarez, R. K., K. C. Welch, et al. (2009). "Flight muscle enzymes and metabolic flux rates during hovering flight of the nectar bat, *Glossophaga soricina*: further evidence of convergence with hummingbirds." *Comparative Biochemistry and Physiology A—Molecular & Integrative Physiology* 153: 136–140.

Tavares, V. D. (2013). "Phyllostomid bat wings from Atlantic Forest ensembles: an ecomorphological study." *Chiroptera Neotropical* 19: 57–70.

Taylor, M. and M. D. Tuttle (2019). *Bats: an Illustrated Guide to All Species*. Washington, Smithsonian Books.

Teeling, E. C. (2009). "Hear, hear: the convergent evolution of echolocation in bats?" *Trends in Ecology & Evolution* 24: 351–354.

Teeling, E. C., M. S. Springer, et al. (2005). "A molecular phylogeny for bats illuminates biogeography and the fossil record." *Science* 307: 580–584.

Temeles, E. J., J. S. Miller, et al. (2010). "Evolution of sexual dimorphism of bill size and shape of hermit hummingbirds (Phaethornithinae): a role for ecological causation." *Philosophical Transactions of the Royal Society B* 365: 1053–1063.

ten Cate, C. (2021). "Re-evaluating vocal production learning in non-oscine birds." *Proceedings of the Royal Society B-Biological Sciences* 376: 1–13.

Thiagavel, J., S. Brinklov, et al. (2020). Sensory and cognitive ecology. *Phyllostomid Bats: a Unique Mammalian Radiation*. T. H. Fleming, L. M. Davalos and M. A. R. Mello. Chicago, University of Chicago Press: 187–204.

Thiagavel, J., C. Cechetto, et al. (2018). "Auditory opportunity and visual constraint enabled the evolution of echolocation in bats." *Nature Communications* 9: 1–10.

Thomas, A. L. R. (1997). "On the tails of birds." *Bioscience* 47: 215–225.

Thornhill, A. H., M. D. Crisp, et al. (2019). "A dated molecular perspective of eucalypt taxonomy, evolution, and diversification." *Australian Systematic Botany* 32: 29–48.

Torres-Vanages, F., A. S. Hadley, et al. (2019). "The landscape genetic signature of pollination by trapliners: evidence from the tropical herb, *Heliconia tortuosa*." *Frontiers in Genetics* 10: 1–12.

Tracy, T. T. and M. C. Baker (1990). "Geographic variation in syllables of house finch songs." *Auk* 116: 666–676.

Traveset, A., J. M. Olesen, et al. (2015). "Bird-flower visitation networks in the Galapagos unveil a widespread interaction release." *Nature Communications* 6(6376).

Tripp, E. A. and P. S. Manos (2008). "Is floral specialization an evolutionary dead-end? Pollination system transitions in *Ruellia* (Acanthaceae)." *Evolution* 62(7): 1712–1736.

Tschapka, M. (2004). "Energy density patterns of nectar resources permit coexistence within a guild of Neotropical flower-visiting bats." *Journal of Zoology* 263: 7–21.

Tyrell, L., B. Goller, et al. (2018). "The orientation of visual space from the perspective of hummingbirds." *Frontiers in Neuroscience* 12: 16.

Upham, N. S., J. A. Esselstyn, et al. (2019). "Inferring the mammal tree: Species-level sets of phylogenies for questions in ecology, evolution, and conservation." *Plos Biology* 17(12).

Vaughan, T. N., J. M. Ryan, et al. (2000). *Mammalogy*. Fort Worth, Texas, Saunders College Publishing.

Vejdani, H. R., D. B. Boerman, et al. (2019). "The dynamics of hovering flight in hummingbirds, insects, and bats with implications for aerial robotics." *Bioinspiration and Biomimetics* 14.

Venable, G. X., K. Gahm, and R. O. Prum (2021). "Hummingbird plumage color diversity exceeds the known gamut of all other birds." *Communications Biology* 5.

Veselka, N., D. D. McErlain, et al. (2010). "A bony connection signals laryngeal echolocation in bats." *Nature* 463: 939–942.

Voigt, C. C., D. H. Kelm, et al. (2006). "Field metabolic rates of phytophagous bats: do pollination strategies of plants make life of nectar-feeders spin faster?" *Journal of Comparative Physiology B-Biochemical Systemic and Environmental Physiology* 176(3): 213–222.

Voigt, C. C. and J. R. Speakman (2007). "Nectar-feeding bats fuel their high metabolism directly with exogenous carbohydrates." *Functional Ecology* 21(5): 913–921.

Voigt, C. C. and Y. Winter (1999). "Energetic cost of hovering flight in nectar-feeding bats (Phyllostomidae: Glossophaginae) and its scaling in moths, birds and bats." *Journal of Comparative Physiology B-Biochemical Systemic and Environmental Physiology* 169(1): 38–48.

Voigt, C. C., A. Zubaid, et al. (2011). "Sources of assimilated proteins in Old and New World phytophagous bats." *Biotropica* 43: 108–113.

von Helversen, U. and H.-U. Reyer (1984). "Nectar intake and energy expenditure in a flower visiting bat." *Oecologia* 63: 178–184.

Voskamp, A., D. J. Baker, et al. (2017). "Global patterns in the divergence between phylogenetic diversity and species richness in terrestrial birds." *Journal of Biogeography* 44: 709–721.

Wang, Y., H. Jiao, et al. (2019). "Functional divergence of bitter taste receptors in a nectar-feeding bird." *Biological Letters* 15: 1–6.

Warren, B. H., E. Bermingham, et al. (2003). "Molecular phylogeography reveals island colonization history and diversification of western Indian Ocean sunbirds (Nectarinia: Nectariniidae)." *Molecular Phylogenetics and Evolution* 29(1): 67–85.

Warrick, D. R., T. Hedrick, et al. (2012). "Hummingbird flight." *Current Biology* 22: R472–R477.

Waterhouse, R. D. (1997). "Some observations on the ecology of the rainbow lorikeet *Trichoglossus haematodus* in Oakley, South Sydney." *Corella* 21: 17–24.

Weinstein, B., C. H. Graham, et al. (2017). "The role of environment, dispersal and competition in explaining reduced co-occurrence among related species." *Plos One* 12(11).

Weinstein, B. G. and C. H. Graham (2017). "Persistent bill and corolla matching despite shifting temporal resources in tropical hummingbird-plant interactions." *Ecology Letters* 20: 326–335.

Weinstein, B. G., B. Tinoco, et al. (2014). "Taxonomic, phylogenetic, and trait beta diversity in South American hummingbirds." *American Naturalist* 184.

Welch, K. C. and C. C. W. Chen (2014). "Sugar flux through the flight muscles of hovering vertebrate nectarivores: a review." *Journal of Comparative Physiology B-Biochemical Systemic and Environmental Physiology* 184(8): 945–959.

Welch, K. C., A. M. Myrka, et al. (2018). "The metabolic flexibility of hovering vertebrate nectarivores." *Physiology* 33(2): 127–137.

Welch Jr., K. C., D. L. Altshuler, et al. (2007). "Oxygen consumption rates in hovering hummingbirds reflect substrate-dependent differences in P/O ratios: carbohydrate as a 'premium fuel.'" *Journal of Experimental Biology* 210: 2146–2153.

Wellbrock, A. H. J., L. R. H. Eckhardt, et al. (2022). "Cool birds: first evidence of energy-saving nocturnal torpor in free-living common swifts *Apus apus* resting in their nests." *Biological Letters* 18.

Wessinger, C. A., M. D. Rausher, et al. (2019). "Adaptation to hummingbird pollination is associated with reduced diversification in *Penstemon*." *Evolution Letters* 3–5: 521–533.

Wester, P. (2014). "Feeding on the wing: hovering in nectar-drinking Old World birds—more common than expected." *Emu* 114: 171–183.

Wethington, S. M., G. C. West, et al. (2005). *Hummingbird Conservation: Discovering Diversity Patterns in Southwest U.S.A.*. Connecting Mountain Islands and Desert Seas: Biodiversity and Management of the Madrean Archipelago II, Tucson, AZ, U.S. Forest Service, Rocky Mountain Research Station.

White, N. D., Z. A. Batz, et al. (2021). "A novel exome set captures phototransduction genes across birds (Aves) enabling efficient analysis of vision evolution." *Molecular Ecology Resources* 2021.

White, N. D. and M. J. Braun (2019). "Extracting phylogenetic signal from phylogenomic data: higher-level relationships of the nightbirds (Strisores)." *Molecular Phylogenetics and Evolution* 141.

Wiles, G. J. and A. P. Brooke (2009). Conservation threats to bats in the tropical Pacific islands and insular Southeast Asia. *Island Bats: Evolution, Ecology, and Conservation*. T. H. Fleming and P. Racey. Chicago, University of Chicago Press: 405–459.

Wilkinson, G. S., D. M. Adams, et al. (2021). "DNA methylation predicts age and provides insight into exceptional longevity of bats." *Nature Communications* 2021: 1–13.

Winkelmann, J. R., F. J. Bonaccorso, et al. (2003). "Home range and territoriality in the least blossom bat, *Macroglossus minimus*, in Papua New Guinea." *Journal of Mammalogy* 84: 561–570.

Wolf, B. O., A. E. McKechnie, et al. (2020). "Extreme and variable torpor among high-elevation Andean hummingbird species." *Biological Letters* 16.

Wooller, R. D., K. C. Richardson, et al. (1988). "The digestion of pollen by some Australian birds." *Australian Journal of Zoology* 36: 357–362.

Workman, R. E., A. M. Myrka, et al. (2018). "Single-molecule, full-length transcript sequencing provides insight into the extreme metabolism of the ruby-throated hummingbird, *Archilochus colibris*." *GigaScience* 7: 1–12.

Yanega, G. M. and M. A. Rubega (2004). "Hummingbird jaw bends to aid insect capture." *Nature* 428: 615.

Yong, E. (2022). *An Immense World, How Animal Senses Reveal the Hidden Realms around Us*. New York, Random House.

Xiang, W.-Q., P. L. Malabrigo, Jr., et al. (2022). "Limited-distance pollen dispersal and low paternal diversity in a bird-pollinated, self-incompatible tree." *Frontiers in Plant Science* 13: 1–12.

Xiong, J., W. M. Fischer, et al. (2000). "Molecular evidence for the early evolution of photosynthesis." *Science* 284: 1724–1730.

Zenata, T. B., B. Dalsgaard, et al. (2017). "Global patterns of interaction specialization in bird-flower networks." *Journal of Biogeography* 44: 1891–1910.

Zenzal, T. J., Jr., F. R. Moore, et al. (2018). "Migratory hummingbirds make their own rules: the decision to resume migration along a barrier." *Animal Behaviour* 137: 215–224.

Zhang, G., C. Cowled, et al. (2013). "Comparative analysis of bat genomes provides insight into the evolution of flight and immunity." *Science* 339: 456–460.

Zhao, Z., Z.-E. Hou, and Li (2021). "Cenozoic Tethyan changes dominated Eurasian animal evolution and diversity patterns." *Zoological Research* 43: 3–13.

Zimmer, C. and D. J. Emlen (2013). *Evolution: Making Sense of Life*. Greenwood Village, CO: Roberts and Company.

INDEX

ABOUT THE AUTHOR

Theodore H. Fleming spent thirty-nine years in academia at the University of Missouri-St. Louis and the University of Miami. He taught ecology courses there and conducted research primarily on plant-visiting bats and their food plants in Panama, Costa Rica, Australia, Mexico, and Arizona. He is currently studying the behavior of the nectar-feeding bat *Leptonycteris yerbabuenae* in southern Arizona and is a world authority on the biology of plant-visiting phyllostomid bats.